George Johnston Allman

GREEK GEOMETRY FROM THALES TO EUCLID

Elibron Classics
www.elibron.com

Elibron Classics series.

ISBN 1-4021-9780-2 (paperback)
ISBN 1-4021-3455-X (hardcover)

This Elibron Classics Replica Edition is an unabridged facsimile of the edition published in 1889 by Hodges, Figgis, & Co., Dublin.

ARCHYTAS.

'non sordidus auctor
naturae verique.'

DUBLIN UNIVERSITY PRESS SERIES.

GREEK GEOMETRY

FROM

THALES TO EUCLID.

BY

GEORGE JOHNSTON ALLMAN, LL.D., D.Sc.;

FELLOW OF THE ROYAL SOCIETY;

PROFESSOR OF MATHEMATICS IN QUEEN'S COLLEGE, GALWAY;

MEMBER OF THE SENATE OF THE ROYAL UNIVERSITY OF IRELAND.

DUBLIN: HODGES, FIGGIS, & CO., GRAFTON-STREET.

LONDON: LONGMANS, GREEN, & CO., PARTERNOSTER-ROW.

1889.

DUBLIN:
PRINTED AT THE UNIVERSITY PRESS,
BY PONSONBY AND WELDRICK.

PREFACE.

THIS book is the reproduction of a Paper, the several parts of which appeared in *Hermathena* during the last eleven years. The favourable reception which from the first it met with on the part of many competent authorities, as well in this country as on the Continent, and the desire which has been expressed in several quarters that the Articles should be collected and published in a volume, have led to this publication.

I have prefixed headings to the chapters, and introduced some additional diagrams. I have also added some notes and an index. Some changes, too, were necessitated by the new form of the Work; and I have made a few corrections, which are indicated for the most part by brackets. With these exceptions the book is textually the same as the Paper in *Hermathena*. In this Paper great pains were taken to ensure accuracy in the references: these I have since checked, and I trust that they will now be found quite reliable.

It has been, throughout, my aim to state clearly the facts as known to us from the original sources, and to make a distinct separation between them and conjectures, however probable the latter might be.

The bust in the frontispiece is taken from Gronovius, *Thesaurus Graecarum Antiquitatum*, Vol. II., Tab. 49. The inscription under it in the engraving is :—

ARCHYTAS

Pythagoricus Mechanicis Clarus
Ex Nummo aereo apud Fulvium Ursinum.

Ample references are given in the notes to the authors whose works I have studied.

It only remains for me now to express my warmest thanks and acknowledgments, in the first place, to my friend DR. JOHN K. INGRAM, Senior Fellow of Trinity College, Dublin, to whom this Work from its inception and during its course is much indebted. Indeed it would scarcely have been written but for the hospitable reception afforded to it in the pages of *Hermathena*, which periodical, edited by DR. INGRAM, enabled me to publish the results of my labours gradually. In the midst of his many and arduous duties, and of his own important literary work, he has been always ready to assist me by his kind encouragement and sound judgment.

I have, in the next place, to acknowledge my great obligations to my late friend and colleague Dr. John F. Davies, Professor of Latin in Queen's College, Galway, whose recent death I deplore. In the later Articles in *Hermathena* I was much aided by his counsel and valuable suggestions: he kindly supervised all the translations that were not purely mathematical; he carefully revised the proofs, and added some critical notes. Nor can I close this reference to Dr. Davies without dwelling for a moment on his rare qualifications as a scholar, his disinterested love of learning, and the nobleness of his personal character.

In conclusion, I have to express my thanks to the Provost and Senior Fellows of Trinity College, Dublin, for including this Work in the Dublin University Press Series.

GEORGE J. ALLMAN.

Queen's College, Galway,
January 10th, 1889.

ERRATA.

Page 49, note 77,	*for*	Dr. h.	*read*	Dr. Ch.
Page 114, note 35,	,,	solution	,,	solutions.
Page 122, note 47,	,,	$\sqrt[3]{\frac{b}{-}}$	,,	$\sqrt[3]{\frac{b}{a}}$.
Page 140, line 23,	,,	discreet	,,	discrete.

CONTENTS.

INTRODUCTION.

CHAPTER I.

THALES.

CHAPTER II.

PYTHAGORAS AND HIS SCHOOL.

CHAPTER III.

THE GEOMETERS OF THE FIFTH CENTURY B.C.— HIPPOCRATES OF CHIOS: DEMOCRITUS.

CHAPTER IV.

ARCHYTAS.

CHAPTER V.

EUDOXUS.

CHAPTER VI.

THE SUCCESSORS OF EUDOXUS—I. MENAECHMUS.

CHAPTER VII.

THE SUCCESSORS OF EUDOXUS—II. DEINOSTRATUS.

CHAPTER VIII.

THE SUCCESSORS OF EUDOXUS—III. ARISTAEUS.

CHAPTER IX.

THEAETETUS.

GREEK GEOMETRY

FROM

THALES TO EUCLID.[1]

INTRODUCTION.

Object of this Work.—Authorities on the Early History of Geometry.—The Historic Summary of Proclus.

IN studying the development of Greek Science, two periods must be carefully distinguished.

The founders of Greek philosophy—Thales and Pythagoras—were also the founders of Greek Science, and from the time of Thales to that of Euclid and the foundation of the Museum of Alexandria, the development of science was, for the most part, the work of the Greek *philosophers*. With the foundation of the School of Alexandria, a second period commences; and henceforth, until the end of the scientific evolution of Greece, the cultivation of science

[1] It has been frequently observed, and is indeed generally admitted, that the present century is characterised by the importance which is attached to historical researches, and by a widely diffused taste for the philosophy of history.

In Mathematics, we have evidence of these prevailing views and tastes in two distinct ways:—

1°. The publication of many recent works on the history of Mathematics, *e.g.*—

Arneth, A., *die Geschichte der reinen Mathematik*, Stuttgart, 1852; *Bretschneider, C. A., *die Geometrie und die Geometer vor Euklides*, Leipzig, was separated from that of philosophy, and pursued for its own sake.

In this work I propose to give some account of the progress of geometry during the first of these periods, and also to notice briefly the chief organs of its development.

For authorities on the early history of geometry we are dependent on scattered notices in ancient writers, many of which have been taken from a work which has unfortunately been lost—the *History of Geometry* by Eudemus of Rhodes, one of the principal pupils of Aristotle. A summary of the history of geometry during the whole period of which I am about to treat has been preserved by Proclus, who most probably derived it from the work of Eudemus. I give it here at length, because I shall frequently have occasion to refer to it in the following pages.

After attributing the origin of geometry to the Egyptians, who, according to the old story, were obliged to invent it in order to restore the landmarks which had been destroyed by the inundation of the Nile, and observing that it is by no means strange that the invention of the sciences should have originated in practical needs, and that,

1870; Suter, H., *Geschichte der mathematischen Wissenschaften* (1st Part), Zurich, 1873; *Hankel, H., *zur Geschichte der Mathematik in Alterthum und Mittel-alter*, Leipzig, 1874 (a posthumous work); *Hoefer, F., *Histoire des Mathématiques*, Paris, 1874. (This forms the fifth volume by M. Hoefer on the history of the sciences, all being parts of the *Histoire Universelle*, published under the direction of M. Duruy.) In studying this subject, I have made use of the works marked thus *. Though the work of M. Hoefer is too metaphysical and is not free from inadvertencies and even errors, yet I have derived advantage from the part which concerns Pythagoras and his ideas. Hankel's book contains some fragments of a great work on the History of Mathematics, which was interrupted by the death of the author. The part treating of the mathematics of the Greeks during the first period—from Thales to the foundation of the School of Alexandria—is fortunately complete. This is an excellent work, and is in many parts distinguished by its depth and originality.

The monograph of M. Bretschneider is most valuable, and is greatly in advance of all that preceded it on the origin of geometry amongst the Greeks. He has collected with great care, and has set out in the original, the fragments further, the transition from sensual perception to reflection, and from that to knowledge, is to be expected, Proclus goes on to say that—

Thales, having visited Egypt, first brought this knowledge into Greece; that he discovered many things himself, and communicated the beginnings of many to his successors, some of which he attempted in a more abstract manner (καθολικώτερον), and some in a more intuitional or sensible manner (αἰσθητικώτερον).

After him, Ameristus [or Mamercus], brother of the poet Stesichorus, is mentioned as celebrated for his zeal in the study of geometry; and Hippias of Elis has recorded that he gained some reputation for his geometry.

After them Pythagoras changed it into the form of a liberal science, regarding its principles in a purely abstract manner, and investigated its theorems from the immaterial and intellectual point of view (ἀΰλως καὶ νοερῶς); he also discovered the theory of incommensurable quantities (τῶν ἀλόγων πραγματείαν), and the construction of the mundane figures [the regular solids].

After him, Anaxagoras of Clazomenae contributed much to geometry, as also did Oenopides of Chios, who was somewhat junior to Anaxagoras; and Plato has

relating to it, which are scattered in ancient writers: I have derived much aid from these citations.

2°. New editions of ancient Mathematical works, some of which had become extremely scarce, *e. g.*—

Theodosii, *Sphaericorum libri tres*, Nizze, Berolini, 1852; Nicomachi Geraseni, *Introductionis Arithmeticae*, lib. II., Hoche, Lipsiae, 1866 (Teubner); Boetii *de Inst. Arithm.*, *&c.*, ed. G. Friedlein, Lipsiae, 1867 (Teubner); Procli Diadochi *in primum Euclidis Elementorum librum commentarii*, ex recog. G. Friedlein, Lipsiae, 1873 (Teubner); Heronis Alexandrini *Geometricorum et Stereometricorum Reliquiae e libris manuscriptis*, edidit F. Hultsch, Berolini, 1864; Pappi Alexandrini *Collectiones quae supersunt e libris manuscriptis Latina interpretatione et commentariis* instruxit F. Hultsch, vol. I., Berolini, 1876: vol. II., *ib.*, 1877.

Occasional portions only of the Greek text of Pappus had been published at various times (see De Morgan in Dr. W. Smith's *Dictionary of Biography*). An Oxford edition, uniform with the great editions of Euclid, Apollonius, and Archimedes, published in the last century, has been long looked for.

mentioned them in his "Rivals" as having won fame by their mathematics.

After these, Hippocrates of Chios, who found the quadrature of the lune, and Theodorus of Cyrene became famous in geometry. Of those whose names have come down to us, Hippocrates is the first writer of Elements.

Plato, who lived after them, contributed to the progress of geometry, and of the other mathematical sciences, through his study of these subjects, and through the mathematical matter introduced in his writings. Amongst his contemporaries were Leodamas of Thasos, Archytas of Tarentum, and Theaetetus of Athens, by all of whom theorems were added or placed on a more scientific basis.

To Leodamas succeeded Neocleides, and his pupil was Leon, who added much to what had been done before. Leon also composed Elements, which, both in regard to the number and the value of the propositions proved, are put together more carefully; he also invented that part of the solution of a problem called its determination (διορισμός)—a test for determining when the problem is possible and when impossible.

Eudoxus of Cnidus, a little younger than Leon and a companion of Plato's pupils, in the first place increased the number of general theorems, added three proportions to the three already existing, and also developed further the things begun by Plato concerning the section,[1] making use, for the purpose, of the analytical method (ταῖς ἀναλύσεσιν).

Amyclas of Heraclea, one of Plato's companions, and Menaechmus, a pupil of Eudoxus and also an associate

[1] Does this mean the cutting of a straight line in extreme and mean ratio, "*sectio aurea*"? or is the reference to the invention of the conic sections? Most probably the former. In *Euclid's Elements*, lib. XIII., the terms *analysis* and *synthesis* are first used and defined by him in connection with theorems relating to the cutting of a line in extreme and mean ratio. See Bretschneider, *die Geometrie vor Euklides*, p. 168.

of Plato, and his brother, Deinostratus, made the whole of geometry more perfect. Theudius of Magnesia appears to have been distinguished in mathematics, as well as in other branches of philosophy, for he made an excellent arrangement of the Elements, and generalized many particular propositions. Athenaeus of Cyzicus [or Cyzicinus of Athens] about the same time became famous in other mathematical studies, but especially in geometry. All these frequented the Academy, and made their researches in common.

Hermotimus of Colophon developed further what had been done by Eudoxus and Theaetetus, discovered much of the Elements, and wrote something on Loci. Philippus Mendaeus [Medmaeus], a pupil of Plato, and drawn by him to mathematical studies, made researches under Plato's direction, and occupied himself with whatever he thought would advance the Platonic philosophy. Thus far those who have written on the history of geometry bring the development of the science.[3]

Proclus goes on to say, Euclid was not much younger than these; he collected the Elements, arranged much of what Eudoxus had discovered, and completed much that had been commenced by Theaetetus; further, he substituted incontrovertible proofs for the lax demonstrations of his predecessors. He lived in the times of the first Ptolemy, by whom, it is said, he was asked whether there was a shorter way to the knowledge of geometry than by his Elements, to which he replied that there was no *royal road* to geometry. Euclid then was younger than the disciples of Plato, but elder than Eratosthenes and Archimedes—who were contemporaries—the latter of whom mentions him. He was of the Platonic sect, and familiar

[3] From these words we infer that the *History of Geometry* by Eudemus is most probably referred to, inasmuch as he lived at the time here indicated, and his history is elsewhere mentioned by Proclus.—Proclus, ed. G. Friedlein, pp. 299, 333, 352, and 379.

with its philosophy, whence also he proposed to himself the construction of the so-called Platonic bodies [the regular solids] as the final aim of his systematisation of the Elements.[4]

[4] Procli Diadochi *in primum Euclidis Elementorum librum commentarii.* Ex recognitione G. Friedlein. Lipsiae, 1873, pp. 64–68.

CHAPTER I.

THALES.

The founder of Greek Geometry.—Characteristic feature of his work.—Distinction between Greek Science and the Science of the Orientals.—Notices of the geometrical work of Thales.—Inferences from these notices as to his geometrical knowledge.—Importance of his work.—The further progress of Geometry was not due to his successors in the Ionic School.

THE first name, then, which meets us in the history of Greek mathematics is that of THALES of Miletus (640–546 B.C.). He lived at the time when his native city, and Ionia in general, were in a flourishing condition, and when an active trade was carried on with Egypt. Thales himself was engaged in trade, is said to have resided in Egypt, and, on his return to Miletus in his old age, to have brought with him from that country the knowledge of geometry and astronomy.

To the knowledge thus introduced he added the capital creation of the geometry of lines, which was essentially abstract in its character. The only geometry known to the Egyptian priests was that of surfaces, together with a sketch of that of solids, a geometry consisting of some simple quadratures and elementary cubatures, which they had obtained empirically; Thales, on the other hand, introduced *abstract* geometry, the object of which is to establish precise *relations* between the different parts of a figure, so that some of them could be found by means of others in a manner strictly rigorous. This was a phenomenon quite new in the world, and due, in fact, to the abstract spirit of the Greeks. In connection with the new impulse given to geometry, there arose with Thales, moreover, scientific astronomy, also an abstract science, and undoubtedly a Greek creation. The astronomy of the Greeks differs from that of the Orientals in this respect, that the astronomy of the latter, which is altogether concrete and empirical, consisted merely in determining the duration of some periods, or in indicating, by means of a mechanical process, the motions of the sun and planets, whilst the astronomy of the Greeks aimed at the discovery of the geometric laws of the motions of the heavenly bodies.[1]

The following notices of the geometrical work of Thales have been preserved:—

(*a*). "He is reported to have first demonstrated that the circle was bisected by its diameter";[2]

(*b*). He is said first to have stated the theorem that the angles at the base of every isosceles triangle are equal, "or, as in archaic fashion he phrased it, *like* (ὁμοῖαι)";[3]

(*c*). Eudemus attributes to him the theorem that when two straight lines cut each other, the vertically opposite angles are equal;[4]

(*d*). "Pamphila[5] relates that he, having learned geometry from the Egyptians, was the first person to describe a right-angled triangle in a circle; others, however, of whom Apollodorus, the calculator (ὁ λογιστικός), is one, say the same of Pythagoras";[6]

(*e*). "He never had any teacher except during the time when he went to Egypt and associated with the priests. Hieronymus also says that he measured the pyramids, making an observation on our shadows when they are of

[1] The importance, for the present research, of bearing in mind this abstract character of Greek science consists in this, that it furnishes a clue by means of which we can, in many cases, recognise theorems of purely Greek growth, and distinguish them from those of eastern extraction. The neglect of this consideration has led some recent writers on the early history of geometry greatly to exaggerate the obligations of the Greeks to the Orientals; whilst others have attributed to the Greeks the discovery of truths which were known to the Egyptians. See, in relation to the distinction between abstract and concrete science, and its bearing on the history of Greek Mathematics, amongst many passages in the works of Auguste Comte, *Système de Politique Positive*, vol. III., ch. iv., p. 297, *sq.*, and vol. I., ch. i., pp. 424–437; and see, also, *les Grands Types de l'Humanité*, par P. Laffitte, vol. II., Leçon 15ième, p. 280, *sq.*—*Appréciation de la Science Antique.*

the same length as ourselves, and applying it to the pyramids."[7] To the same effect Pliny—" Mensuram altitudinis earum omnemque similem deprehendere invenit Thales Milesius, umbram metiendo, qua hora par esse corpori solet;"[8]

(This is told in a different manner by Plutarch. Niloxenus is introduced as conversing with Thales concerning Amasis, King of Egypt.—" Although he [Amasis] admired you [Thales] for other things, yet he particularly liked the manner by which you measured the height of the pyramid without any trouble or instrument; for, by merely placing a staff at the extremity of the shadow which the pyramid casts, you formed two triangles by the contact of the sunbeams, and showed that the height of the pyramid was to the length of the staff in the same ratio as their respective shadows").[9]

(*f*). Proclus tells us that Thales measured the distance of vessels from the shore by a geometrical process, and that Eudemus, in his history of geometry, refers the theorem Eucl. I. 26 to Thales, for he says that it is necessary to use this theorem in determining the distance of ships at sea according to the method employed by Thales in this investigation;[10]

(*g*). Proclus, or rather Eudemus, tells us in the passage quoted above *in extenso* that Thales brought the knowledge of geometry to Greece, and added many things,

[2] Proclus, ed. Friedlein, p. 157.

[3] *Ibid*, p. 250.

[4] *Ibid*, p. 299.

[5] Pamphila was a female historian who lived at the time of Nero; an Epidaurian according to Suidas; an Egyptian according to Photius.

[6] Diogenes Laertius, I., c. i., n. 3, ed. C. G. Cobet, p. 6.

[7] *ὁ δὲ Ἱερώνυμος καὶ ἐκμετρῆσαί φησιν αὐτὸν τὰς πυραμίδας ἐκ τῆς σκιᾶς παρατηρήσαντα ὅτε ἡμῖν ἰσομεγέθεις εἰσί.* Diog. Laert., I., c. i., n. 6, ed. Cobet, p. 6.

[8] Plin. *Nat. Hist.* XXXVI., 17.

[9] Plut. *Sept. Sap. Conviv.* 2. vol. III., p. 174, ed. Didot.

[10] Proclus, ed. Friedlein, p. 352.

attempting some in a more abstract manner, and some in a more intuitional or sensible manner.[11]

Let us now examine what inferences as to the geometrical knowledge of Thales can be drawn from the preceding notices.

First inference.—Thales must have known the theorem that the sum of the three angles of a triangle is equal to two right angles.

Pamphila, in (*d*), refers to the discovery of the property of a circle that all triangles described on a diameter as base, with their vertices on the circumference, have their vertical angles right.[12]

Assuming, then, that this theorem was known to Thales, he must have known that the sum of the three angles of any right-angled triangle is equal to two right angles; for, if the vertex of any of these right-angled triangles be connected with the centre of the circle, the right-angled triangle will be resolved into two isosceles triangles; and since the angles at the base of an isosceles triangle are equal—a theorem attributed to Thales (*b*)—it follows that the sum of the angles at the base of the right-angled triangle is equal to the vertical angle, and that therefore the

[11] Proclus, ed. Friedlein, p. 65.

[12] This is unquestionably the discovery referred to. The manner in which it has been stated by Diogenes Laertius shows that he did not distinguish between a problem and a theorem; and further that he was ignorant of geometry. To this effect Proclus:—"When, therefore, anyone proposes to inscribe an equilateral triangle in a circle he proposes a problem; for it is possible to inscribe one that is not equilateral. But when anyone asserts that the angles at the base of an isosceles triangle are equal, he must affirm that he proposes a theorem; for it is not possible that the angles at the base of an isosceles triangle should be unequal to each other. On which account if anyone, stating it as a problem, should say that he wishes to inscribe a right angle in a semicircle, he must be considered as ignorant of geometry, since every angle in a semicircle is necessarily a right one."—Taylor's *Proclus*, vol. I., p. 110. Procl. ed. Friedlein, pp. 79, 80.

Sir G. C. Lewis has subjected himself to the same criticism when he says—"According to Pamphila, he first solved the problem of inscribing a right-angled triangle in a circle."—G. Cornewall Lewis, *Historical Survey of the Astronomy of the Ancients*, p. 83.

sum of the three angles of the right-angled triangle is equal to two right angles. Further, since any triangle can be resolved into two right-angled triangles, it follows immediately that the sum of the three angles of any triangle is equal to two right angles. If, then, we accept the evidence of Pamphila as satisfactory, we are forced to the conclusion that Thales must have known this theorem. No doubt the knowledge of this theorem (Euclid I., 32) is required in the proof given in the Elements of Euclid of the property of the circle (III., 31), the discovery of which is attributed to Thales by Pamphila, and some writers have inferred hence

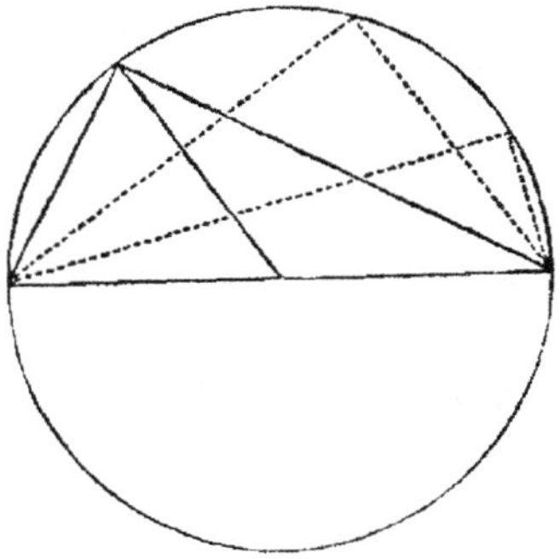

that Thales must have known the theorem (I., 32).[13] Although I agree with this conclusion, for the reasons given above, yet I consider the inference founded on the demonstration given by Euclid to be inadmissible, for we are informed by Proclus, on the authority of Eudemus, that the theorem (Euclid I., 32) was first proved in a general way by the Pythagoreans, and their proof, which does not differ substantially from that given by Euclid, has been preserved by Proclus.[14] Further, Geminus states that the ancient geometers observed the equality to two right angles in each species of triangle separately, first in equilateral, then in isosceles, and lastly in scalene triangles,[15] and it is plain

[13] See P. Laffitte, *op. cit.*, vol. II., p. 291. Cf. F. A. Finger, *de Primordiis Geometriae apud Graecos*, p. 20, Heidelbergae, 1831.

[14] Proclus, ed. Friedlein, p. 379.

[15] Apollonii *Conica*, ed. Halleius, p. 9, Oxon. 1710.

that the geometers older than the Pythagoreans can be no other than Thales and his successors in the Ionic school.

If I may be permitted to offer a conjecture, in conformity with the notice of Geminus, as to the manner in which the theorem was arrived at in the different species of triangles, I would suggest that Thales had been led by the concrete geometry of the Egyptians to contemplate floors covered with tiles in the form of equilateral triangles or regular hexagons,[16] and had observed that six equilateral triangles could be placed round a common vertex; from which he saw that six such angles made up four right angles, and that consequently the sum of the three angles of an equilateral triangle is equal to two right angles (*c*). The observation of a floor covered with square tiles would lead to a similiar conclusion with respect to the isosceles right-angled triangle.[17] Further, if a perpendicular be drawn from a vertex of an equilateral triangle on the opposite side,[18] the triangle is divided into two right-angled triangles, which are in every respect equal to each other, hence the sum of the three angles of each of these right-angled triangles is easily seen to be two right angles. If now we suppose that Thales was led to examine whether the property, which he had observed in two distinct kinds of right-angled triangles, held generally for all right-angled triangles, it seems to me that, by com-

[16] Floors or walls covered with tiles of various colours were common in Egypt. See Wilkinson's *Ancient Egyptians*, vol. II., pp. 287 and 292.

[17] Athough the theorem that "only three kinds of regular polygons—the equilateral triangle, the square, and the hexagon—can be placed about a point so as to fill a space," is attributed by Proclus to Pythagoras or his school (*ἔστι τὸ θεώρημα τοῦτο Πυθαγόρειον*: Proclus, ed. Friedlein, p. 305), yet it is difficult to conceive that the Egyptians—who erected the pyramids—had not a practical knowledge of the fact that tiles of the forms above mentioned could be placed so as to form a continuous plane surface.

[18] Though we are informed by Proclus (ed. Friedlein, p. 283), that Oenopides of Chios first investigated (*ἐζήτησεν*) this problem, yet Thales, and indeed the Egyptians, who were furnished with the square, could not be ignorant of its mechanical solution. Observe that we are expressly told by Proclus that Thales attempted some things in an intuitional or sensible manner.

pleting the rectangle and drawing the second diagonal, he could easily see that the diagonals are equal, that they bisect each other, and that the vertical angle of the right-angled triangle is equal to the sum of the base angles. Further, if he constructed several right-angled triangles on the same hypotenuse he could see that their vertices are all equally distant from the middle point of their common hypotenuse, and therefore lie on the circumference of a circle described on that line as diameter, which is the theorem in question. It may be noticed that this remarkable property of the circle, with which, in fact, abstract geometry was inaugurated, struck the imagination of Dante:—

> "O se del mezzo cerchio far si puote
> Triangol sì, ch'un retto non avesse."
>
> *Par.* c. xiii. 101.

Second inference.—The conception of geometrical loci is due to Thales.

We are informed by Eudemus (*f*) that Thales knew that a triangle is determined if its base and base angles are given; further, we have seen that Thales knew that, if the base is given, and the base angles not given separately, but their sum known to be a right angle, then there could be described an unlimited number of triangles satisfying the conditions of the question, and that their vertices all lie on the circumference of a circle described on the base as diameter. Hence it is manifest that the important conception of *geometrical loci*, which is attributed by Montucla, and after him by Chasles and other writers on the History of Mathematics, to the school of Plato,[19] had been formed by Thales.

[19] Montucla, *Histoire des Mathématiques*, Tome I., p. 183, Paris, 1758. Chasles, *Aperçu historique des Méthodes en Géométrie*, p. 5, Bruxelles, 1837. Chasles in the history of geometry before Euclid copies Montucla, and we have a remarkable instance of this here, for Chasles, after Montucla, calls Plato "ce chef du *Lycée*."

Third inference.—Thales discovered the theorem that the sides of equiangular triangles are proportional.

The knowledge of this theorem is distinctly attributed to Thales by Plutarch in a passage quoted above (*e*). On the other hand, Hieronymus of Rhodes, a pupil of Aristotle, according to the testimony of Diogenes Laertius,[20] says that Thales measured the height of the pyramids by watching when bodies cast shadows of their own length, and to the same effect Pliny in the passage quoted above (*e*). Bretschneider thinks that Plutarch has spun out the story told by Hieronymus, attributing to Thales the knowledge of his own times; denies to Thales the knowledge of the theorem in question, and says that there is no trace of any theorems concerning similarity before Pythagoras.[21] He says, further, that the Egyptians were altogether ignorant of the doctrine of the similarity of figures, that we do not find amongst them any trace of the doctrine of proportion, and that Greek writers say that this part of their mathematical knowledge was derived from the Babylonians or Chaldaeans.[22] Bretschneider also endeavours to show that Thales could have obtained the solution of the second practical problem—the determination of the distance of a ship from the shore—by geometrical construction, a method long before known to the Egyptians.[23] Now, as Bretschneider denies to the Egyptians and to Thales any knowledge of the doctrine of proportion, it was plainly necessary, on this supposition, that Thales should find a sufficient extent of free and level ground on which to construct a triangle of the same dimensions as that he wished to measure; and even if he could have found such ground, the great length of the sides would have rendered the operations very diffi-

[20] But we have seen that the account given by Diogenes Laertius of the discovery of Thales mentioned by Pamphila is unintelligible, and evinces ignorance of geometry on his part.

[21] Bretsch. *die Geometrie und Geometer vor Euklides*, pp. 45, 46.

[22] *Ibid*, p. 18.

[23] *Ibid*, pp. 43, 44.

cult.[24] It is much simpler to accept the testimony of Plutarch, and suppose that the method of superseding such operations by using similar triangles is due to Thales.

If Thales had employed a right-angled triangle,[25] he could have solved this problem by the same principle which, we are told by Plutarch, he used in measuring the height of the pyramid, the only difference being that the right-angled triangle is in one case in a vertical, and in the other in a horizontal plane.

From what has been said, it is plain that there is a natural connection between the several theorems attributed to Thales, and that the two practical applications which he made of his geometrical knowledge are also connected with each other.

Let us now proceed to consider the importance of the work of Thales :—

I. In a scientific point of view :—

(*a*). We see, in the first place, that by his two theorems he founded the geometry of lines, which has ever since remained the principal part of geometry.[26]

Vainly do some recent writers refer these geometrical discoveries of Thales to the Egyptians; in doing so they ignore the distinction between the geometry of lines, which

[24] In reference to this I may quote the following passage from Clairaut, *Élémens de Géométrie*, pp. 34–35. Paris, 1741.

"La méthode qu'on vient de donner pour mesurer les terrains, dans lesquels on ne sçauroit tirer de lignes, fait souvent naître de grandes difficultés dans la pratique. On trouve rarement un espace uni et libre, assez grand pour faire des triangles égaux à ceux du terrain dont on cherche la mesure. Et même quand on en trouveroit, la grande longueur des côtés des triangles pourroit rendre les opérations très-difficiles : abaisser une perpendiculaire sur une ligne du point qui en est éloigné seulement de 500 toises, ce seroit un ouvrage extrêmement pénible, et peut-être impracticable. Il importe donc d'avoir un moyen qui supplée à ces grandes opérations. Ce moyen s'offre comme de lui-même. Il vient," &c.

[25] Observe that the inventions of the square and level are attributed by Pliny (*Nat. Hist.*, VII., 57) to Theodorus of Samos, who was a contemporary of Thales. They were, however, known long before this period to the Egyptians; so that to Theodorus is due at most the honour of having introduced them into Greece.

[26] Auguste Comte, *Système de Politique Positive*, vol. III., p. 297.

we owe to the genius of the Greeks, and that of areas and volumes—the only geometry known, and that empirically, to the ancient priesthoods. This view is confirmed by an ancient papyrus, that of Rhind,[27] which is now in the British Museum. It contains a complete applied mathematics, in which the measurement of figures and solids plays the principal part; there are no theorems properly so called; everything is stated in the form of problems, not in general terms but in distinct numbers, *e.g.*—to measure a rectangle the sides of which contain two and ten units of length; to find the surface of a circular area whose diameter is six units; to mark out in a field a right-angled triangle whose sides measure ten and four units; to describe a trapezium whose parallel sides are six and four units, and each of the other sides twenty units. We find also in it indications for the measurement of solids, particularly of pyramids, whole and truncated.

It appears from the above that the Egyptians had made great progress in practical geometry. Of their proficiency and skill in geometrical constructions we have also the direct testimony of the ancients; for example, Democritus says: "No one has ever excelled me in the construction of lines according to certain indications—not even the so-called Egyptian Harpedonaptae."[28]

(*b*). Thales may, in the second place, be fairly considered to have laid the foundation of Algebra, for his first theorem establishes an equation in the true sense of the word, while the second institutes a proportion.[29]

[27] Birch, in Lepsius' *Zeitschrift für Aegyptische Sprache und Alterthumskunde* (1868, p. 108). Bretschneider, *Geometrie vor Euklides*, p. 16. F. Hoefer, *Histoire des Mathématiques*, p. 69. Since this Paper was sent to the press, Dr. August Eisenlohr, of Heidelberg, has published this papyrus with a translation and commentary under the title "*ein mathematisches Handbuch der alten Ægypter*, Leipzig, 1877."

[28] Mullach, *Fragmenta Philosophorum Graecorum*, p. 371, Democritus ap. Clem. Alex. *Strom.* I., p. 357, ed. Potter.

[29] Auguste Comte, *Système de Politique Positive*, vol. III., p. 300.

II. In a philosophic point of view:—

We see that in these two theorems of Thales the first type of a *natural law*—*i.e.* the expression of a fixed dependence between different quantities, or, in another form, the disentanglement of constancy in the midst of variety—has decisively arisen.[30]

III. Lastly, in a practical point of view:—

Thales furnished the first example of an application of theoretical geometry to practice,[31] and laid the foundation of an important branch of the same—the measurement of heights and distances.

I have now pointed out the importance of the geometrical discoveries of Thales, and attempted to appreciate his work. His successors of the Ionic School followed him in other lines of thought, and were, for the most part, occupied with physical theories on the nature of the universe—speculations which have their representatives at the present time—and added little or nothing to the development of science, except in astronomy. The further progress of geometry was certainly not due to them.

Without doubt Anaxagoras of Clazomenae, one of the latest representatives of this School, is said to have been occupied during his exile with the problem of the quadrature of the circle; but this was in his old age, and after the works of another School—to which the early progress of geometry was really due—had become the common property of the Hellenic race. I refer to the immortal School of Pythagoras.

[30] P. Laffitte, *les Grands Types de l'Humanité*, vol. II., p. 292.
[31] *Ibid.*, p. 294.

CHAPTER II.

PYTHAGORAS AND HIS SCHOOL.

State of Hellas about the middle of the sixth century, B.C.—Pythagoras, probable date of his birth and death.—Difficulties in treating of Pythagoras and the early Pythagoreans.—Pythagoras first raised mathematics to the rank of a science, and added two new branches, Arithmetic and Music.—Notices of the geometrical work of this School.—It is much concerned with the Geometry of Areas, and is Egyptian in its character.—Generation of Squares, Gnomon, Gnomonic Numbers.—Pythagorean Triangles.—The Theorem of the Three Squares.—Construction of Regular Polygons and of the Regular Solids.—Discovery of Incommensurable Quantities.—The Application of Areas.—The Doctrine of Proportion and of the Similarity of Figures.—Theorems erroneously attributed to Pythagoras and his School.—Conclusions from the foregoing examination.—Estimate of the state of Geometry *circ.* 480 B.C.—The Theory of Proportion.—The Ancients regarded Proportion not merely as a branch of Arithmetic but as the bond of Mathematics.—Estimate of the services of Pythagoras.

ABOUT the middle of the sixth century before the Christian era a great change had taken place: Ionia, no longer free and prosperous, had fallen under the yoke, first of Lydia, then of Persia, and the very name Ionian—the name by which the Greeks were known in the whole East—had become a reproach, and was shunned by their kinsmen on the other side of the Aegean.[1] On the other hand, Athens and Sparta had not become pre-eminent; the days of Marathon and Salamis were yet to come. Meanwhile the glory of the Hellenic name was maintained chiefly by the Italic Greeks, who were then in the height of their prosperity, and had recently obtained for their territory the well-earned appellation of ἡ μεγάλη Ἑλλάς.[2] It should be noted, too, that at this period there was great commercial intercourse between the Hellenic cities of Italy and Asia; and further, that some of them, as Sybaris and Miletus on

[1] Herodotus, I., 143. [2] Polybius, II., 39.

the one hand, and Tarentum and Cnidus on the other, were bound by ties of the most intimate character.[3] It is not surprising, then, that after the Persian conquest of Ionia, Pythagoras, Xenophanes, and others, left their native country, and, following the current of civilisation, removed to Magna Graecia.

As the introduction of geometry into Greece is by common consent attributed to Thales, so all[4] are agreed that to PYTHAGORAS of Samos, the second of the great philosophers of Greece, and founder of the Italic School, is due the honour of having raised mathematics to the rank of a science.

The statements of ancient writers concerning this great man are most conflicting, and all that relates to him personally is involved in obscurity: for example, the dates given for his birth vary within the limits of eighty-four years—43rd to 64th Olympiad.[5] It seems desirable, however, if for no other reason than to fix our ideas, that we should adopt some definite date for the birth of Pythagoras; and there is an additional reason for doing so, inasmuch as some writers, by neglecting this, have become confused, and fallen into inconsistencies in the notices which they have given of his life. Of the various dates which have been assigned for the birth of Pythagoras, the one which seems to me to harmonise best with the records of the most trustworthy writers is that given by Ritter, and adopted by Grote, Brandis, Ueberweg, and Hankel, namely, about 580 B. C. (49th Olymp.) This date would accord with the following statements:—

That Pythagoras had personal relations with Thales, then old, of whom he was regarded by all antiquity as the

[3] Herod., VI., 21, and III., 138.

[4] Aristotle, Diogenes Laertius, Proclus, amongst others.

[5] See G. H. Lewes, *Biographical History of Philosophy*, Book II., c. ii., where the various dates given by scholars are cited.

successor, and by whom he was incited to visit Egypt,[6]—mother of all the civilisation of the West;

That he left his country being still a young man, and, on this supposition as to the date of his birth, in the early years of the reign of Croesus (560–546 B.C.), when Ionia was still free;

That he resided in Egypt many years, so that he learned the Egyptian language, and became imbued with the philosophy of the priests of the country;[7]

That he probably visited Crete and Tyre, and may have even extended his journeys to Babylon, at that time Chaldaean and free;

That on his return to Samos, finding his country under the tyranny of Polycrates,[8] and Ionia under the dominion of the Persians, he migrated to Italy in the early years of Tarquinius Superbus;[9]

And that he founded his Brotherhood at Crotona, where for the space of twenty years or more he lived and taught, being held in the highest estimation, and even looked on almost as divine by the population—native as well as Hellenic; and then, soon after the destruction of Sybaris (510 B.C.), being banished by a democratic party under Cylon, he removed to Metapontum, where he died soon afterwards.

All who have treated of Pythagoras and the Pythagoreans have experienced great difficulties. These difficulties are due partly to the circumstance that the reports of the earlier and most reliable authorities have for the most part been lost, while those which have come down to us are not always consistent with each other. On the other hand, we have pretty full accounts from later writers, especially those of the Neo-Pythagorean School; but these notices, which

[6] Iamblichus, *Vit. Pyth.*, c. ii., 12.

[7] Isocrates is the oldest authority for this, *Busiris*, c. ii.

[8] Diog. Laert., VIII., c. i., 3; Aristoxenus, ap. Porphyr., *Vit. Pyth.*, 9.

[9] Cicero, *de Rep.* II., 15; *Tusc. Disp.*, I., xvi., 38.

are mixed up with fables, were written with a particular object in view, and are in general highly coloured; they are particularly to be suspected, as Zeller has remarked, because the notices are fuller and more circumstantial the greater the interval from Pythagoras. Some recent authors, therefore, even go to the length of omitting from their account of the Pythagoreans everything which depends solely on the evidence of the Neo-Pythagoreans. In doing so, these authors no doubt effect a simplification, but it seems to me that they are not justified in this proceeding, as the Neo-Pythagoreans had access to ancient and reliable authorities which have unfortunately been lost since.[10]

Though the difficulties to which I refer have been felt chiefly by those who have treated of the Pythagorean *philosophy*, yet we cannot, in the present inquiry, altogether escape from them; for, in the first place, there was, in the whole period of which we treat, an intimate connection between the growth of philosophy and that of science, each re-acting on the other; and, further, this was particularly the case in the School of Pythagoras, owing to the fact, that whilst on the one hand he united the study of geometry with that of arithmetic, on the other he made numbers the base of his philosophical system, as well physical as metaphysical.

It is to be observed, too, that the early Pythagoreans published nothing, and that, moreover, with a noble self-denial, they referred back to their master all their discoveries. Hence, it is not possible to separate what was done by him from what was done by his early disciples, and we are under the necessity, therefore, of treating the work of the early Pythagorean School as a whole.[11]

[10] For example, the *History of Geometry*, by Eudemus of Rhodes, one of the principal pupils of Aristotle, is quoted by Theon of Smyrna, Proclus, Simplicius, and Eutocius, the last two of whom lived in the reign of Justinian. Eudemus also wrote a *History of Astronomy*. Theophrastus, too, Aristotle's successor, wrote *Histories of Arithmetic, Geometry, and Astronomy.*

[11] "Pythagoras and his earliest successors do not appear to have committed any

All agree, as was stated above, that Pythagoras first raised mathematics to the rank of a science, and that we owe to him two new branches—arithmetic and music.

We have the following statements on the subject:—

(1) "In the age of these Philosophers [the Eleats and Atomists], and even before them, lived those called Pythagoreans, who first applied themselves to mathematics, a science they improved: and, penetrated with it, they fancied that the principles of mathematics were the principles of all things";[12]

(2) Eudemus informs us, in the passage quoted above *in extenso*, that "Pythagoras changed geometry into the form of a liberal science, regarding its principles in a purely abstract manner, and investigated his theorems from the immaterial and intellectual point of view"; and that "he also discovered the theory of irrational quantities, and the construction of the mundane figures [the five regular solids]";[13]

(3) "It was Pythagoras, also, who carried geometry to perfection, after Moeris[14] had first found out the principles of the elements of that science, as Anticleides tells us in the second book of his *History of Alexander;* and the part of the science to which Pythagoras applied himself above all others was arithmetic";[15]

(4) Pythagoras seems to have esteemed arithmetic above

of their doctrines to writing. According to Porphyrius (*Vit. Pyth.* p. 40) Lysis and Archippus collected in a written form some of the principal Pythagorean doctrines, which were handed down as heir-looms in their families, under strict injunctions that they should not be made public. But amid the different and inconsistent accounts of the matter, the first publication of the Pythagorean doctrines is pretty uniformly attributed to Philolaus."—Smith's *Dictionary, in v. Philolaus.* Philolaus was born at Crotona, or at Tarentum, and was a contemporary of Socrates and Democritus. See Diog. Laert., *Vit. Pyth.*, VIII., c. i., 15; *Vit. Empedoclis*, VIII., c. ii., 2; and *Vit. Democriti*, IX., c. vii., 6. See also Iamblichus, *Vit. Pyth.*, c. xviii., 88.

[12] Aristot. *Met.*, I., v., 985^b, 23, ed. Bekker.

[13] Proclus, ed. Friedlein, p. 65.

[14] An ancient king of Egypt, who lived 900 years before Herodotus.

[15] Diog. Laert., VIII., c. i., 11, ed. Cobet, p. 207.

everything, and to have advanced it by diverting it from the service of commerce, and likening all things to numbers;[16]

(5) He was the first person who introduced measures and weights among the Greeks, as Aristoxenus the musician informs us;[17]

(6) He discovered the numerical relations of the musical scale;[18]

(7) The word *mathematics* originated with the Pythagoreans;[19]

(8) The Pythagoreans made a four-fold division of mathematical science, attributing one of its parts to the how many, τὸ πόσον, and the other to the how much, τὸ πηλίκον; and they assigned to each of these parts a two-fold division. For they said that discrete quantity, or the *how many*, either subsists by itself, or must be considered with relation to some other; but that continued quantity, or the *how much*, is either stable or in motion. Hence they affirmed that arithmetic contemplates that discrete quantity which subsists by itself, but music that which is related to another; and that geometry considers continued quantity so far as it is immovable; but astronomy (τὴν σφαιρικὴν) contemplates continued quantity so far as it is of a self-motive nature;[20]

(9) Favorinus says that he employed definitions on account of [*i. e.* arising out of] the mathematical subjects to which he applied himself (ὅροις χρήσασθαι διὰ τῆς μαθηματικῆς ὕλης).[21]

[16] Aristoxenus, *Fragm.* ap. Stob. *Eclog. Phys.*, I., ii., 6; ed. Heeren, vol. I., p. 17.

[17] Diog. Laert., VIII., c. i., 13, ed. Cobet, p. 208.

[18] τόν τε κανόνα τὸν ἐκ μιᾶς χορδῆς εὑρεῖν. Diog. Laert., VIII., c. i., 11, ed. Cobet, p. 207.

[19] Proclus, ed. Friedlein, p. 45.

[20] *Ibid.*, p. 35. As to the distinction between τὸ πηλίκον, continuous, and τὸ πόσον, discrete, quantity, see Iambl., *in Nicomachi Geraseni Arithmeticam introductionem*, ed. Tennulius, p. 148.

[21] Diog. Laert., VIII., c. i., 25, ed. Cobet, p. 215.

As to the particular work done by this school in geometry, the following statements have been handed down to us :—

(*a*) The Pythagoreans define a point as "unity having position (μονάδα προσλαβοῦσαν θέσιν)";[22]

(*b*) They considered a point as analogous to the monad, a line to the duad, a superficies to the triad, and a body to the tetrad;[23]

(*c*) The plane around a point is completely filled by six equilateral triangles, four squares, or three regular hexagons: this is a Pythagorean theorem;[24]

(*d*) The Peripatetic Eudemus ascribes to the Pythagoreans the discovery of the theorem that the interior angles of a triangle are equal to two right angles (Eucl. I. 32), and states their method of proving it, which was substantially the same as that of Euclid;[25]

(*e*) Proclus informs us in his commentary on Euclid I., 44, that Eudemus says that the problems concerning the application of areas—in which the term "application" is not to be taken in its restricted sense (παραβολή), in which it is used in this proposition, but also in its wider signification, embracing ὑπερβολὴ and ἔλλειψις, in which it is used in the 28th and 29th propositions of the Sixth Book—are old, and inventions of the Pythagoreans;[26]

(*f*) This is to some extent confirmed by Plutarch, who

[22] Proclus, ed. Friedlein, p. 95.

[23] *Ibid.*, p. 97.

[24] *Ibid.*, p. 305.

[25] *Ibid.*, p. 379.

[26] *Ibid.*, p. 419. The words of Proclus are interesting :—

"According to Eudemus, the inventions respecting the *application*, *excess*, and *defect* of areas are ancient (ἀρχαῖα), and are due to the Pythagoreans. Moderns borrowing these names transferred them to the so-called conic lines—the parabola, the hyperbola, the ellipse; as the older school in their nomenclature concerning the description of areas *in plano* on a finite right line regarded the terms thus :—

"An area is said to be *applied* (παραβάλλειν) to a given right line when an area equal in content to some given one is described thereon; but when the base of the area is greater than the given line, then the area is said to be in *excess* (ὑπερβάλλειν);

says that Pythagoras sacrificed an ox on account of the geometrical diagram, as Apollodotus [-rus] says:—

ἡνίκα Πυθαγόρης τὸ περικλεὲς εὕρετο γράμμα,
κεῖν' ἐφ' ὅτῳ λαμπρὴν ἤγετο βουθυσίην,

either the one relating to the hypotenuse—namely, that the square on it is equal to the sum of the squares on the sides—or that relating to the problem concerning the application of areas (εἴτε πρόβλημα περὶ τοῦ χωρίου τῆς παραβολῆς);[27]

(*g*) "One of the most essentially geometrical (γεωμετρικωτάτοις) theorems, or rather problems, is to construct a figure equal to one and similiar to another given figure, for the solution of which also they say that Pythagoras offered a sacrifice: and indeed it is finer and more elegant than the theorem which shows that the square on the hypotenuse is equal to the sum of the squares on the sides";[28]

(*h*) Eudemus, in the passage already quoted from Proclus, says: Pythagoras discovered the construction of the regular solids;[29]

(*i*) "But particularly as to Hippasus, who was a Pythagorean, they say that he perished in the sea on account of his impiety, inasmuch as he boasted that he first divulged the knowledge of the sphere with the twelve pentagons

but when the base is less, so that some part of the given line lies without the described area, then the area is said to be in *defect* (ἐλλείπειν). Euclid uses in this way, in his Sixth Book, the terms *excess* and *defect*. . . . The term *application* (παραβάλλειν), which we owe to the Pythagoreans, has this signification."

27 Plutarch, *non posse suaviter vivi sec. Epicurum*. c. xi.; Plut., *Opera*, ed. Didot, vol. IV., p. 1338. Some authors, rendering περὶ τοῦ χωρίου τῆς παραβολῆς "concerning the area of the parabola," have ascribed to Pythagoras the quadrature of the parabola—which was in fact one of the great discoveries of Archimedes; and this, though Archimedes himself tells us that no one before him had considered the question; and though further he gives in his letter to Dositheus the history of his discovery, which, as is well known, was first obtained from mechanical considerations, and then by geometrical reasonings.

28 Plutarch, *Symp.*, VIII., *Quaestio* 2, c. iv. Plut. *Opera*, ed. Didot, vol. IV., p. 877.

29 Proclus, ed. Friedlein, p. 65.

[the ordinate dodecahedron inscribed in the sphere]: Hippasus assumed the glory of the discovery to himself, whereas everything belonged to Him—for thus they designate Pythagoras, and do not call him by name";[30]

(*j*) The triple interwoven triangle or Pentagram—star-shaped regular pentagon—was used as a symbol or sign of

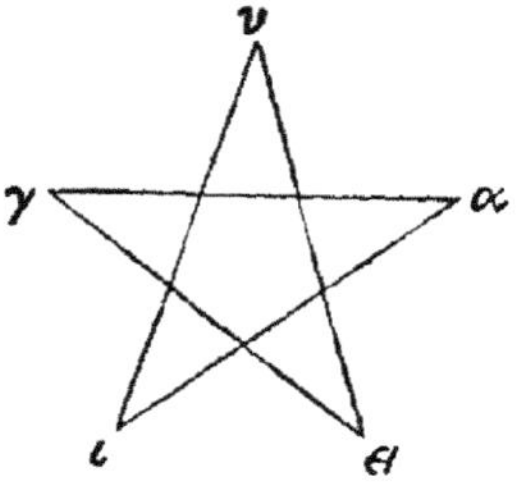

recognition by the Pythagoreans, and was called by them Health (ὑγιεία);[31]

(*k*) The discovery of the law of the three squares (Eucl. I., 47), commonly called the *Theorem of Pythagoras*, is attributed to him by—amongst others—Vitruvius,[32] Diogenes Laertius,[33] Proclus,[34] and Plutarch (*f*). Plutarch, however, attributes to the Egyptians the knowledge of this theorem in the particular case where the sides are 3, 4, and 5;[35]

(*l*) One of the methods of finding right-angled tri-

[30] Iambl., *Vit. Pyth.*, c. xviii., 88.

[31] Scholiast on Aristophanes, *Nub.* 609; also Lucian, *pro Lapsu in Salut.*, s. 5, vol. I., pp. 447, 8; ed. C. Jacobitz. That the Pythagoreans used such symbols we learn from Iamblichus (*Vit. Pyth.*, c. xxxiii., 237 and 238). This figure is referred to Pythagoras himself, and in the middle ages was called *Pythagorae figura*. It is said to have obtained its special name from his having written the letters υ, γ, ι, θ (= ει), α, at its prominent vertices. We learn from Kepler (*Opera Omnia*, ed. Frisch, vol. V., p. 122) that even so late as Paracelsus it was regarded by him as the symbol of health. See Chasles, *Histoire de Géométrie*, pp. 477, *sq.*

[32] *De Arch.*, IX., cap. ii.

[33] Where the same couplet from Apollodorus as that in (*f*) is found, except that κλεινὴν ἤγαγε occurs in place of λαμπρὴν ἤγετο. Diog. Laert., VIII., c. i., 11, p. 207, ed. Cobet.

[34] Proclus, p. 426, ed. Friedlein.

[35] *de Is. et Osir.*, c. 56. Plut. *Op.*, vol. III., p. 457, Didot.

angles whose sides can be expressed in numbers—that setting out from the odd numbers—is attributed to Pythagoras;[36]

(*m*) The discovery of irrational quantities is ascribed to Pythagoras by Eudemus in the passage quoted above from Proclus;[37]

(*n*) The three proportions—arithmetical, geometrical, and harmonical—were known to Pythagoras;[38]

(*o*) "Formerly, in the time of Pythagoras and the mathematicians under him, there were three means only—the arithmetical, the geometrical, and the third in order which was known by the name *ὑπεναντία*, but which Archytas and Hippasus designated the harmonical, since it appeared to include the ratios concerning harmony and melody (*μετακληθεῖσα ὅτι τοὺς κατὰ τὸ ἁρμοσμένον καὶ ἐμμελὲς ἐφαίνετο λόγους περιέχουσα*)";[39]

(*p*) With reference to the means corresponding to these proportions, Iamblichus says:[40]—"We must now speak of the most perfect proportion, consisting of four terms, and properly called the musical, for it clearly contains the musical ratios of harmonical symphonies. It is said to be an invention of the Babylonians; and to have been first brought into Greece by Pythagoras";[41]

[36] Proclus, ed. Friedlein, p. 428; Heronis Alex., *Geom. et Ster. Rel.*, ed. F. Hultsch, pp. 56, 146.

[37] Proclus, ed. Friedlein, p. 65.

[38] Nicom. G. *Introd. Ar.*, c. xxii., ed. R. Hoche, p. 122.

[39] Iamblichus, *in Nicom. Arithm.*, ed. Ten., p. 141.

[40] *Ibid.*, p. 167.

[41] As an example of this proportion, Nicomachus gives the numbers 6, 8, 9, 12, the harmonical and arithmetical means between two numbers forming a geometrical proportion with the numbers themselves. (Nicom. *Instit. Arithm.*, ed. Ast., p. 153, and *Animad.*, p. 329: see also Iambl., *in Nicom. Arithm.*, ed. Ten., p. 172, *sq.*)

Hankel, commenting on this passage of Iamblichus, says: "What we are to do with the report, that this proportion was known to the Babylonians, and only brought into Greece by Pythagoras, must be left to the judgment of the reader."—*Geschichte der Mathematik*, p. 105. In another part of his book, however, after referring to two authentic documents of the Babylonians which have come down to us, he says: "We cannot, therefore, doubt that the Babylonians occupied (*q*) The doctrine of arithmetical progressions is attributed to [the School of] Pythagoras ;[42]

(*r*) It would appear that he had considered the special case of *triangular* numbers. Thus Lucian :—ΠΥΘ. Εἶτ' ἐπὶ τουτεοῖσιν ἀριθμέειν. ΑΓ. Οἶδα καὶ νῦν ἀριθμεῖν. ΠΥΘ. Πῶς ἀριθμέεις; ΑΓ. Ἕν, δύο, τρία, τέτταρα. ΠΥΘ. Ὁρᾷς; ἃ σὺ δοκέεις τέτταρα, ταῦτα δέκα ἐστὶ καὶ τρίγωνον ἐντελὲς καὶ ἡμέτερον ὅρκιον.[43]

(*s*) Another of his doctrines was, that of all solid figures the sphere was the most beautiful ; and of all plane figures, the circle.[44]

(*t*) "Also Iamblichus, in his commentary on the Categories of Aristotle, says that Aristotle may perhaps not have squared the circle; but that the Pythagoreans had done so, as is evident, he adds, from the demonstrations of the Pythagorean Sextus, who had got by tradition the manner of proof."[45]

On examining the purely geometrical work of Pythagoras and his early disciples, we observe that it is much concerned with the geometry of areas, and we are indeed struck with its Egyptian character. This appears in the theorem (*c*) concerning the filling up a plane by regular polygons, as already noted; in the construction of the regular solids (*h*), for some of them are found in the Egyptian architecture; in the problems concerning the application of areas (*e*); and, lastly, in the law of the three

themselves with such progressions [arithmetrical and geometrical]; and a Greek notice that they knew proportions, nay, even invented the so-called perfect or musical proportion, gains thereby in value."—*Ibid.*, p. 67.

[42] *Theologumena Arithmetica*, p. 61, ed. F. Ast, Lipsiae, 1817.

[43] PYTH. Then I will teach you to count. BUYER. I know how to count already. PYTH. How do you count? BUYER. One, two, three, four. PYTH. Do you see? What you take to be four, that is ten and a perfect triangle and our oath. Lucian, Βίων πρᾶσις, 4, vol. I., p. 317, ed. Jacobitz.

[44] Καὶ τῶν σχημάτων τὸ κάλλιστον σφαῖραν εἶναι τῶν στερεῶν, τῶν δ' ἐπιπέδων κύκλον, Diog. Laert., *Vit. Pyth.*, VIII., c. i., 19, ed. Cobet, p. 212.

[45] Simplicius, *Comment.*, &c., ap. Bretsch., *Geom. vor Eukl.*, p. 108. [Simplicii, *in Aristotelis Physicorum libros quattuor priores Commentaria*, p. 60, ed. Hermannus Diels, Berolini, 1882.]

squares (*k*), coupled with the rule given by Pythagoras for the construction of right-angled triangles in numbers (*l*).

According to Plutarch, the Egyptians knew that a triangle whose sides consist of 3, 4, and 5 parts, must be right-angled. "The Egyptians may perhaps have imagined the nature of the universe like the most beautiful triangle, as also Plato appears to have made use of it in his work on the State, where he sketches the picture of matrimony. That triangle contains one of the perpendiculars of 3, the base of 4, and the hypotenuse of 5 parts, the square of which is equal to those of the containing sides. The perpendicular may be regarded as the male, the base as the female, the hypotenuse as the offspring of both, and thus Osiris as the originating principle (ἀρχὴ), Isis as the receptive principle (ὑποδοχή), and Horus as the product (ἀποτέλεσμα)." [46]

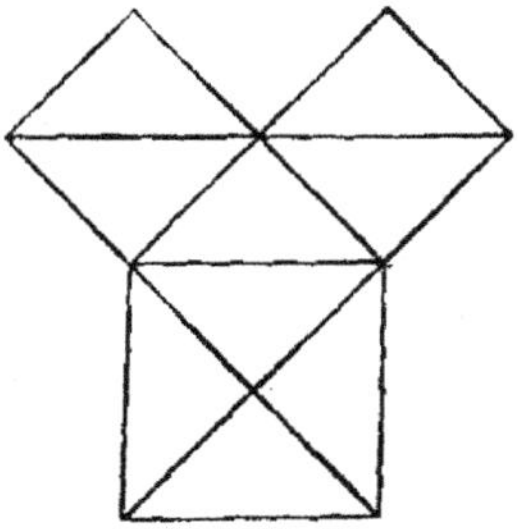

This passage is remarkable, and seems to indicate the way in which the knowledge of the useful geometrical fact enunciated in it may have been arrived at by the Egyptians. The contemplation of a draught-board, or of a floor covered with square tiles, or of a wall ruled with squares,[47] would at once show that the square constructed

[46] Plutarch, *de Is et Osir.*, c. 56, vol. III., p. 457, ed. Didot.

[47] It was the custom of the Egyptians, where a subject was to be drawn, to rule the walls of the building accurately with squares before the figures were introduced. See Wilkinson's *Ancient Egyptians*, vol. II., pp. 265, 267.

on the diagonal of a square is equal to the sum of the squares constructed on the sides—each containing four of the right-angled isosceles triangles into which one of the squares is divided by its diagonal.

Although this observation would not serve them for practical uses, on account of the impossibility of presenting it arithmetically, yet it must have shown the possibility of constructing a square which would be the sum of two squares, and encouraged them to attempt the solution of the problem numerically. Now, the Egyptians, with whom speculations concerning generation were in vogue, could scarcely fail to have perceived, from the observation of a

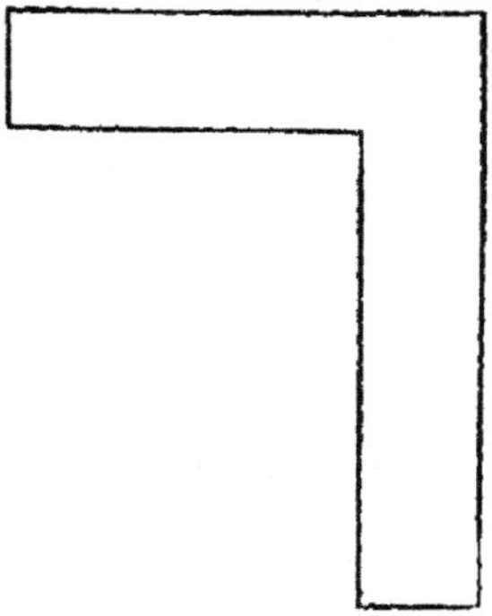

chequered board, that the element in the successive formation of squares is the gnomon (*γνώμων*)[48], or common

[48] Γνώμων means that by which anything is known, or criterion; its oldest concrete signification seems to be the carpenter's square (*norma*), by which a right angle is known. Hence, it came to denote a perpendicular, of which, indeed, it was the archaic name, as we learn from Proclus on Euclid, I., 12:—Τοῦτο τὸ πρόβλημα πρῶτον Οἰνοπίδης ἐζήτησεν χρήσιμον αὐτὸ πρὸς ἀστρολογίαν οἰόμενος· ὀνομάζει δὲ τὴν κάθετον ἀρχαϊκῶς κατὰ γνώμονα, διότι καὶ ὁ γνώμων πρὸς ὀρθάς ἐστι τῷ ὁρίζοντι (Proclus, ed. Friedlein, p. 283). Gnomon is also an instrument for measuring altitudes, by means of which the meridian can be found; it denotes, further, the index or style of a sundial, the shadow of which points out the hours.

In geometry it means the square or rectangle about the diagonal of a square or rectangle, together with the two complements, on account of the resemblance of the figure to a carpenter's square; and then, more generally, the similar figure carpenter's square, which was known to them.[49] It remained for them only to examine whether some particular gnomon might not be metamorphosed into a square, and, therefore, *vice versâ*. The solution would then be easy, being furnished at once from the contemplation of a floor or board composed of squares.

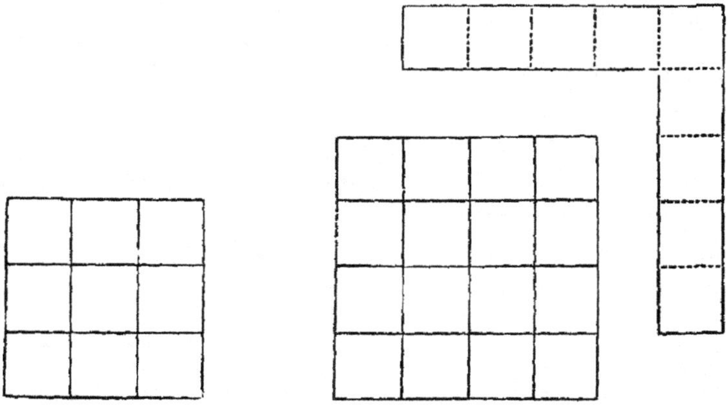

Each gnomon consists of an odd number of squares, and the successive gnomons correspond to the successive odd numbers,[50] and include, therefore, all odd squares. Suppose, now, two squares are given, one consisting of 16 and the other of 9 unit squares, and that it is proposed to form another square out of them. It is plain that the square

with regard to any parallelogram, as defined by Euclid, II., Def. 2. Again, in a still more general signification, it means the figure which, being added to any figure, preserves the original form. See Hero, *Definitiones* (59).

When gnomons are added successively in this manner to a square monad, the first gnomon may be regarded as that consisting of three square monads, and is indeed the constituent of a simple Greek fret; the second, of five square monads, &c.; hence we have the *gnomonic* numbers, which were also looked on as male, or generating.

[49] Wilkinson's *Ancient Egyptians*, vol. II., p. 111.

[50] It may be observed here that we first count with *counters*, as is indicated by the Greek ψηφίζειν and the Latin *calculare*. The counters might be equal squares, as well as any other like objects. There is an indication that the odd numbers were first regarded in this manner in the name *gnomonic* numbers

consisting of 9 unit squares can take the form of the fourth gnomon, which, being placed round the former square, will generate a new square containing 25 unit squares. Similarly, it may have been observed that the 12th gnomon, consisting of 25 unit squares, could be transformed into a square, each of whose sides contains 5 units, and thus it may have been seen conversely that the latter square, by taking the gnomonic, or generating, form with respect to the square on 12 units as base, would produce the square of 13 units, and so on.

This, then, is my attempt to interpret what Plutarch has told us concerning Isis, Osiris, and Horus, bearing in mind that the odd, or gnomonic, numbers were regarded by Pythagoras as male, or generating.[51]

which the Pythagoreans applied to them, and that term was used in the same signification by Aristotle, and by subsequent writers, even up to Kepler. See Arist., *Phys.*, lib. III., ed. Bekker, vol. I., p. 203; Stob., *Eclog.*, ab Heeren, vol. I., p. 24, and note; Kepleri, *Opera Omnia*, ed. Ch. Frisch, vol. VIII., *Mathematica*, pp. 164, *sq.*

[51] This seems to me to throw light on some of the oppositions which are found in the table of principles attributed by Aristotle to certain Pythagoreans (*Metaph.*, I., v., 986ª, ed. Bekker).

The odd—or *gnomonic*—numbers are finite; the even, infinite. Odd numbers were regarded also as male, or generating. Further, by the addition of successive gnonoms—consisting, as we have seen, each of an odd number of units—to the original unit square or monad, the square form is preserved. On the other hand, if we start from the simplest oblong (ἑτερόμηκες), consisting of two unit squares, or monads, in juxtaposition, and place about it, after the manner of a gnomon—

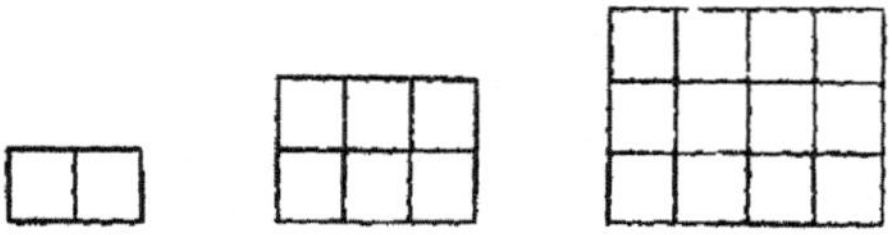

and gnomon, as we have seen, was used in this more extended sense also at a later period—4 unit squares, and then in succession in like manner 6, 8, . . . unit squares, the oblong form ἑτερόμηκες will be preserved. The elements, then, which generate a square are odd, while those of which the oblong is made up are even. The limited, the odd, the male, and the square, occur on one side of the table: while the unlimited, the even, the female, and the oblong, are met with on the other side.

The correctness of this view is confirmed by the following passage preserved by Stobaeus:—Ἔτι δὲ τῇ μονάδι τῶν ἐφεξῆς περισσῶν γνωμόνων περιτιθεμένων, ὁ γινόμενος ἀεὶ τετράγωνός ἐστι. τῶν δὲ ἀρτίων ὁμοίως περιτιθέμενων, ἑτερομήκεις καὶ ἄνισοι πάντες ἀποβαίνουσιν· ἴσον δὲ ἰσάκις οὐδείς.

It is another matter to see that the triangle formed by 3, 4, and 5 units is right-angled, and this I think the Egyptians may have first arrived at by an induction founded on direct measurement, the opportunity for which was furnished to them by their pavements, or chequered plane surfaces.

The method given above for the formation of the square constructed on 5 units as the sum of those constructed on 4 units and on 3 units, and of that constructed on 13 units, as the sum of those constructed on 12 units and 5 units, required only to be generalized in order to

"Explicanda haec sunt ex antiqua Pythagoricorum terminologia. Γνώμονες nempe de quibus hic loquitur auctor, vocabantur apud eos omnes numeri impares, *Jo. Philop. ad Aristot. Phys.*, l. iii., p. 131: Καὶ οἱ ἀριθμητικοὶ δὲ γνώμονας καλοῦσι πάντας τοὺς περιττοὺς ἀριθμούς. Causam adjicit Simplicius ad eundem locum, Γνώμονας δὲ ἐκάλουν τοὺς περιττοὺς οἱ Πυθαγόρειοι διότι προστιθέμενοι τοῖς τετραγώνοις, τὸ αὐτὸ σχῆμα φυλάττουσι, ὥσπερ καὶ οἱ ἐν γεωμετρίᾳ γνώμονες. Quae nostro loco leguntur jam satis clara erunt. Vult nempe auctor, monade addita ad primum gnomonem, ad sequentes autem summam, quam proxime antecedentes numeri efficiunt, semper prodire numeros quadratos, *v. c.* positis gnomonibus 3, 5, 7, 9, primum $1 + 3 = 2^2$, tunc porro $1 + 3$ (*i. e.* 4) $+ 5 = 3^2$, $9 + 7 = 4^2$, $16 + 9 = 5^2$, et sic porro, cf. *Tiedem. Geist der Speculat. Philos.*, pp. 107, 108. Reliqua expedita sunt." Stob. *Eclog.* ab. Heeren, lib. I., p. 24, and note.

The passage of Aristotle referred to is—σημεῖον δ' εἶναι τούτου τὸ συμβαῖνον ἐπὶ τῶν ἀριθμῶν. περιτιθεμένων γὰρ τῶν γνωμόνων περὶ τὸ ἓν καὶ χωρὶς ὁτὲ μὲν ἄλλο ἀεὶ γίγνεσθαι τὸ εἶδος ὁτὲ δὲ ἕν. *Phys.*, III., iv., p. 203[a], 12.

Compare ἀλλ' ἔστι τινὰ αὐξανόμενα ἃ οὐκ ἀλλοιοῦται, οἷον τὸ τετράγωνον γνώμονος περιτεθέντος ηὔξηται μέν, ἀλλοιότερον δὲ οὐδὲν γεγένηται. *Cat.*, XIV., 15[a], 30, Arist., ed. Bekker.

Hankel gives a different explanation of the opposition between the square and oblong :—

"When the Pythagoreans discovered the theory of the Irrational, and recognised its importance, it must, as will be at once admitted, appear most striking that the oppositions, which present themselves so naturally, of Rational and Irrational have no place in their table. Should they not be contained under the image of square and rectangle, which, in the extraction of the square root, have led precisely to those ideas?" *Geschichte der Mathematik*, p. 110, note.

Hankel also says—"Upon what the comparison of the odd with the limited may have been based, and whether upon the theory of the gnomons, can scarcely be made out now." *Ibid.* p. 109, note.

May not the gnomon be looked on as *framing*, as it were, or limiting the squares?

enable Pythagoras to arrive at his rule for finding right-angled triangles, which we are told sets out from the odd numbers.

The two rules of Pythagoras and of Plato are given by Proclus:—"But there are delivered certain methods of finding triangles of this kind [*sc.* right-angled triangles whose sides can be expressed by numbers], one of which they refer to Plato, but the other to Pythagoras, as originating from odd numbers. For Pythagoras places a given odd number as the lesser of the sides about the right angle, and when he has taken the square constructed on it, and diminished it by unity, he places half the remainder as the greater of the sides about the right angle; and when he has added unity to this, he gets the hypotenuse. Thus, for example, when he has taken 3, and has formed from it a square number, and from this number 9 has taken unity, he takes the half of 8, that is 4, and to this again he adds unity, and makes 5; and thus obtains a right-angled triangle, having one of its sides of 3, the other of 4, and the hypotenuse of 5 units. But the Platonic method originates from even numbers. For when he has taken a given even number, he places it as one of the sides about the right angle, and when he has divided this into half, and squared the half, by adding unity to this square he gets the hypotenuse, but by subtracting unity from the square he forms the remaining side about the right angle. Thus, for example, taking 4, and squaring its half, 2, and thus getting 4, then subtracting 1, he gets 3, and by adding 1 he gets 5; and he obtains the same triangle as by the former method."[52] It should be observed, however, that this is not necessarily the case; for example, we may obtain by the method of Plato a triangle whose sides are 8, 15, and 17 units, which cannot be got by the Pythagorean method.

The nth square together with the nth gnomon is the

[52] Proclus, ed. Friedlein, p. 428. Hero, *Geom. et ster. rel.*, ed. Hultsch, pp. 56, 57.

$(n+1)^{th}$ square; if the nth gnomon contains m^2 unit squares, m being an *odd* number, we have

$$2n+1=m^2, \therefore n=\frac{m^2-1}{2};$$

hence the rule of Pythagoras. Similarly the sum of two successive gnomons contains an even number of unit squares and may therefore consist of m^2 unit squares, where m is an *even* number; we have then

$$(2n-1)+(2n+1)=m^2, \text{ or } n=\left(\frac{m}{2}\right)^2:$$

hence the rule ascribed to Plato by Proclus.[53] This passage of Proclus, which is correctly interpreted by Hoefer,[54] was understood by Kepler,[55] who, indeed, was familiar with this work of Proclus, and often quotes it in his *Harmonia Mundi*.

Let us now examine how Pythagoras proved the theorem of the three squares. Though he could have discovered it as a consequence of the theorem concerning the proportionality of the sides of equiangular triangles, attributed above to Thales, yet there is no indication whatever of his having arrived at it in that deductive manner. On the other hand, the proof given in the Elements of Euclid clearly points to such an origin, for it depends on the theorem that the square on a side of a right-angled triangle is equal to the rectangle under the hypotenuse and its adjacent segment made by the perpendicular on it from the right angle —a theorem which follows at once from the similarity

[53] This rule is ascribed to Architas [no doubt, Archytas of Tarentum] by Boetius, *Geom.*, ed. Friedlein, p. 408.

[54] Hoefer, *Histoire des Math.*, p. 112.

[55] Kepleri, *Opera Omnia*, ed. Frisch, vol. VIII., p. 163 et seq. It may be observed that this method is capable of further extension, *e.g.*: the sum of 9 (an odd square number) successive gnomons may contain an odd number (say 49 × 9) of square units; hence we obtain a right-angled triangle in numbers, whose hypotenuse exceeds one side by 9 units—the three sides being 20, 21, and 29. Plato's method may be extended in like manner.

of each of the partial triangles into which the original right-angled triangle is broken up by the perpendicular, with the whole. That the proof in the Elements is not the way in which the theorem was discovered is indeed stated directly by Proclus, who says:—

"If we attend to those who wish to investigate antiquity, we shall find them referring the present theorem to Pythagoras, . . . For my own part, I admire those who first investigated the truth of this theorem: but I admire still more the author of the Elements, because he has not only secured it by evident demonstration, but because he reduced it into a more general theorem in his sixth book by strict reasoning [Euclid, VI., 31]."[56]

The simplest and most natural way of arriving at the theorem is the following, as suggested by Bretschneider[57]:—

A square can be dissected into the sum of two squares and two equal rectangles, as in Euclid, II., 4; these two rectangles can, by drawing their diagonals, be decomposed into four equal right-angled triangles, the sum of the sides

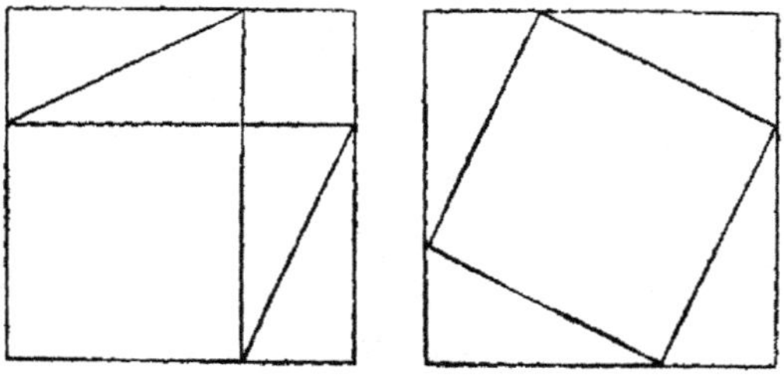

of each being the side of the square: again, these four right-angled triangles can be placed so that a vertex of each shall be in one of the corners of the square in such a way that a greater and less side are in continuation. The original square is thus dissected into the four triangles as

[56] Proclus, ed. Friedlein, p. 426.
[57] Bretsch., *Geom. vor Eukl.*, p. 82. This proof is old: see Camerer, *Euclidis Element.*, vol. I., p. 444, and references given there.

before and the figure within, which is the square on the hypotenuse. This square then must be equal to the sum of the squares on the sides of the right-angled triangle.

Hankel, in quoting this proof from Bretschneider, says that it may be objected that it bears by no means a specifically Greek colouring, but reminds us of the Indian method. This hypothesis as to the oriental origin of the theorem seems to me to be well founded. I would, however, attribute the discovery to the Egyptians, inasmuch as the theorem concerns the geometry of areas, and as the method used is that of the dissection of figures, for which the Egyptians were famous, as we have already seen. Moreover, the theorem concerning the areas connected with two lines and their sum (Euclid, II., 4), which admits also of arithmetical interpretation, was certainly within their reach. The gnomon by which any square exceeds another breaks up naturally into a square and two equal rectangles. I think also that the Egyptians knew that the difference between the squares on two lines is equal to the rectangle under their sum and difference—though they would not have stated it in that abstract manner. The two squares may be placed with a common vertex and adjacent sides coinciding in direction, so that their difference is a gnomon. This gnomon can, on account of the equality of the two complements,[68] be transformed into a rectangle which can be constructed by producing the side of the greater square so that it shall be equal to itself, and then we have the figure of Euclid, II., 5, or to the side of the lesser square, in which case we have the figure of Euclid, II., 6. Indeed I have little hesitation in attributing to the Egyptians the contents of the first ten propositions of the second book of Euclid. In the demonstrations of

[68] This theorem (Euclid, I., 43) Bretschneider says was called the "theorem of the gnomon." I do not know of any authority for this statement. If the theorem were so called, the word gnomon was not used in it either as defined by Euclid (II. *Def.* 2), or in the more general signification in Hero (*Def.* 58).

propositions 5, 6, 7, and 8, use is made of the gnomon, and propositions 9 and 10 also can be proved similarly without the aid of Euclid, I., 47.

It is well known that the Pythagoreans were much occupied with the construction of regular polygons and solids, which in their cosmology played an essential part as the fundamental forms of the elements of the universe.[59]

We can trace the origin of these mathematical speculations in the theorem (*c*) that "the plane around a point is completely filled by six equilateral triangles or four squares, or three regular hexagons," a theorem attributed to the Pythagoreans, but which must have been known as a fact to the Egyptians. Plato also makes the Pythagorean Timaeus explain—"Each straight-lined figure consists of triangles, but all triangles can be dissected into rectangular ones, which are either isosceles or scalene. Among the latter the most beautiful is that out of the doubling of which an equilateral arises, or in which the square of the greater perpendicular is three times that of the smaller, or in which the smaller perpendicular is half the hypotenuse.

But two or four right-angled isosceles triangles, properly put together, form the square; two or six of the most beautiful scalene right-angled triangles form the equilateral triangle; and out of these two figures arise the solids which correspond with the four elements of the real world, the tetrahedron, octahedron, icosahedron, and the cube."[60]

[59] Hankel says it cannot be ascertained with precision how far the Pythagoreans had penetrated into this theory, namely, whether the construction of the regular pentagon and ordinate dodecahedron was known to them. Hankel, *Geschichte der Mathematik*, p. 95, note.

[60] Plato, *Tim.*, c. XX., 53, D., *sq.*, vol. VII., ed. Stallbaum, p. 224, *sq.*

This dissection of figures into right-angled triangles may be fairly referred to Pythagoras, and indeed may have been derived by him from the Egyptians.

The construction of the regular solids is distinctly ascribed to Pythagoras himself by Eudemus, in the passage in which he briefly states the principal services of Pythagoras to geometry. Of the five regular solids, three—the tetrahedron, the cube, and the octahedron—were certainly known to the Egyptians, and are to be found in their architecture. There remain, then, the icosahedron and the dodecahedron. Let us now examine what is required for the construction of these two solids.

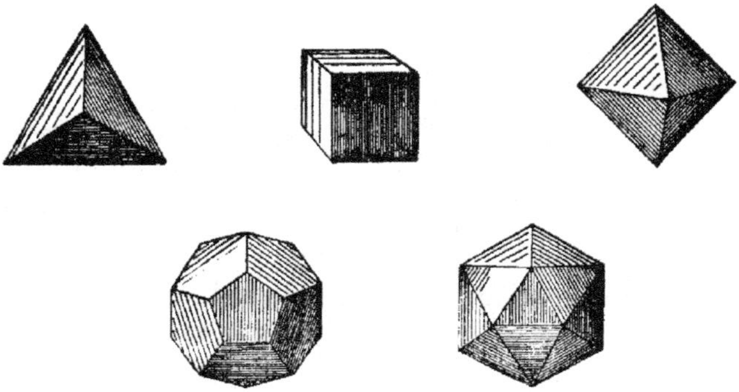

In the formation of the tetrahedron, three, and in that of the octahedron, four, equal equilateral triangles had been placed with a common vertex and adjacent sides coincident, and it was known, too, that if six such triangles were placed round a common vertex with their adjacent sides coincident, they would lie in a plane, and that, therefore, no solid could be formed in that manner from them. It remained, then, to try whether five such equilateral triangles could be placed at a common vertex in like manner: on trial it would be found that they could be so placed, and that their bases would form a regular pentagon. The existence of a regular pentagon would thus be known. It was also known from the formation of the cube that three squares could be placed in a similar way with a common vertex; and that, further, if three equal and regular hexagons were placed round a point as common vertex with adjacent sides coincident, they would form a plane. It remained, then, only to try whether three equal regular pentagons could be placed with a common vertex, and in a similar way; this on trial would be found possible, and would lead to the construction of the regular dodecahedron, which was the regular solid last arrived at.[61]

We see, then, that the construction of the regular pentagon is required for the formation of each of these two regular solids, and that therefore it must have been a discovery of Pythagoras. We have now to examine what knowledge of geometry was required for the solution of this problem.

If any vertex of a regular pentagon be connected with the two remote ones, an isosceles triangle will be formed having each of the base angles double the vertical angle. The construction of the regular pentagon depends, therefore, on the description of such a triangle (Euclid, IV., 10). Now, if either base angle of such a triangle be bisected, the isosceles triangle will be decomposed into two triangles, which are evidently also both isosceles. It is also evident that the one of which the base of the proposed is a side is equiangular with it. From a comparison of the sides of these two triangles it will appear at once by the second theorem, attributed above to Thales, that the problem is reduced to cutting a straight line so that one segment shall be a mean proportional between the whole line and the other segment (Euclid, VI., 30), or so that the rectangle under the whole line and one part shall be equal to the square on the other part (Euclid, II., 11). To effect this, let us suppose the square on the greater segment to be

[61] The four elements had been represented by the four other regular solids; the dodecahedron was then taken symbolically for the universe.

constructed on one side of the line, and the rectangle under the whole line and the lesser segment on the other side. It is evident that by adding to both the rectangle under the whole line and the greater segment, the problem is reduced to the following:—To produce a given straight line so that the rectangle under the whole line thus produced and the part produced shall be equal to the square on the given line, or, in the language of the ancients, to apply to a given straight line a rectangle which shall be equal to a given area—in this case the square on the given line—and which shall be *excessive* by a square. Now it is to be observed that the problem is solved in this manner by Euclid (VI., 30, 1st method), and that we learn from Eudemus that the problems concerning the *application* of areas and their *excess* and *defect* are old, and inventions of the Pythagoreans (*e*).[62]

[62] It may be objected that this reasoning presupposes a knowledge, on the part of Pythagoras, of the method of geometrical analysis, which was invented by Plato more than a century later.

While admitting that it contains the germ of that method, I reply, in the first place, that this manner of reasoning was not only natural and spontaneous, but that in fact in the solution of problems there was no other way of proceeding. And, to anticipate a little, we shall see, secondly, that the oldest fragment of Greek geometry extant—that namely by Hippocrates of Chios—contains traces of an analytical method, and that, moreover, Proclus ascribes to Hippocrates, who, it will appear, was taught by the Pythagoreans, the method of reduction (ἀπαγωγή), a systematisation, as it seems to me, of the manner of reasoning that was spontaneous with Pythagoras. Proclus defines ἀπαγωγή to be "a transition from one problem or theorem to another, which being known or determined, the thing proposed is also plain. For example: when the duplication of the cube is investigated, geometers reduce the question to another to which this is consequent, *i.e.* the finding of two mean proportionals, and afterwards they inquire how between two given straight lines two mean proportionals may be found. But Hippocrates of Chios is reported to have been the first inventor of geometrical reduction (ἀπαγωγή): who also squared the lune, and made many other discoveries in geometry, and who was excelled by no geometer in his powers of construction."—Proclus, ed. Friedlein, p. 212, *sq.* Lastly, we shall find that the passages in Diogenes Laertius and Proclus, which are relied on in support of the statement that Plato invented this method, prove nothing more than that Plato communicated it to Leodamas of Thasos. For my part, I am convinced that the gradual elaboration of this famous method—by which mathematics rose above the Elements—is due to the Pythagorean philosophers from the founder to Theodorus of Cyrene and Archytas of Tarentum, who were Plato's masters in mathematics.

The statements, then, of Iamblichus concerning Hippasus (*i*)—that he divulged the sphere with the twelve pentagons; and of Lucian and the scholiast on Aristophanes (*j*)—that the pentagram was used as a symbol of recognition amongst the Pythagoreans, become of greater importance. We learn, too, from Iamblichus that the Pythagoreans made use of signs for that purpose.[63]

Further, the discovery of irrational magnitudes is ascribed to Pythagoras in the same passage of Eudemus (*m*), and this discovery has been ever regarded as one of the greatest of antiquity. It is commonly assumed that Pythagoras was led to this theory from the consideration of the isosceles right-angled triangle. It seems to me, however, more probable that the discovery of incommensurable magnitudes was rather owing to the problem—To cut a line in extreme and mean ratio. From the solution of this problem it follows at once that, if on the greater segment of a line, so cut, a part be taken equal to the less, the greater segment, regarded as a new line, will be cut in a similar manner; and this process can be continued without end. On the other hand, if a similar method be adopted in the case of any two lines which are capable of numerical representation, the process would end. Hence would arise the distinction between commensurable and incommensurable quantities.

A reference to Euclid, X., 2, will show that the above method is the one used to prove that two magnitudes are incommensurable. And in Euclid, X., 3, it will be seen that the greatest common measure of two commensurable magnitudes is found by this process of continued subtraction.

It seems probable that Pythagoras, to whom is attributed one of the rules for representing the sides of right-angled triangles in numbers, tried to find the sides of an isosceles right-angled triangle numerically, and that, fail-

[63] Iambl., *Vit. Pyth.*, c. cxxxiii., p. 77, ed. Didot.

ing in the attempt, he suspected that the hypotenuse and a side had no common measure. He may have demonstrated the incommensurability of the side of a square and its diagonal. The nature of the old proof—which consisted of a *reductio ad absurdum*, showing that if the diagonal be commensurable with the side, it would follow that the same number would be odd and even[64]—makes it more probable, however, that this was accomplished by his successors.

The existence of the irrational, as well as that of the regular dodecahedron, appears to have been regarded also by the school as one of their chief discoveries, and to have been preserved as a secret. It is remarkable, too, that a story similar to that told by Iamblichus of Hippasus is narrated of the person who first published the idea of the irrational, namely, that he suffered shipwreck, &c.[65]

Eudemus ascribes the problems concerning the application of figures to the Pythagoreans. The simplest cases of the problems (Euclid, VI., 28, 29)—those, namely, in which the given parallelogram is a square—correspond to the problem: To cut a straight line internally or externally so that the rectangle under the segments shall be equal to a given rectilineal figure. On examination it will be found that the solution of these problems depends on the problem Euclid, II., 14, and the theorems Euclid, II., 5 and 6, which we have seen were probably known to the Egyptians, together with the law of the three squares (Euclid, I., 47).

The finding of a mean proportional between two given lines, or the construction of a square which shall be equal

[64] Aristoteles, *Analyt. Prior.*, I., c. xxiii., 41ᵃ, 26, and c. xliv., 50ᵃ, 37, ed. Bekker.

Euclid has preserved this proof, X., 117. Hankel thinks he did so probably for its historical interest only, since the irrationality follows self-evidently from X., 9; and X., 117, is merely an appendix.—Hankel, *Gesch. der Math.*, p. 102, note.

[65] *Untersuchungen über die neu aufgefundenen Scholien des Proklus Diadochus zu Euclid's Elementen*, von Dr. Joachim Heinrich Knoche, Herford, 1865, pp. 20 and 23.

to a given rectangle, must be referred, I have no doubt, to Pythagoras. The rectangle can be easily thrown into the form of a gnomon, and then exhibited as the difference between two squares, and therefore as a square by means of the law of the three squares.

Lastly, the solution of the problem to construct a rectilineal figure which shall be equal to one and similar to another given rectilineal figure is attributed by Plutarch to Pythagoras. The solution of this problem depends on the application of areas, and requires a knowledge of the theorems:—that similar rectilineal figures are to each other as the squares on their homologous sides; that if three lines be in geometrical proportion, the first is to the third as the square on the first is to the square on the second; and also on the solution of the problem, to find a mean proportional between two given straight lines. Now, we shall see later that Hippocrates of Chios—who was instructed in geometry by the Pythagoreans—must have known these theorems and the solution of this problem. We are justified, therefore, in ascribing this theorem also, if not (with Plutarch) to Pythagoras, at least to his early successors.

The theorem that similar polygons are to each other in the duplicate ratio of their homologous sides involves a first sketch, at least, of the doctrine of proportion.

That we owe the foundation and development of the doctrine of proportion to Pythagoras and his disciples is confirmed by the testimony of Nicomachus (*n*) and Iamblichus (*o*) and (*p*).

From these passages it appears that the early Pythagoreans were acquainted not only with the arithmetical and geometrical means between two magnitudes, but also with their harmonical mean, which was then called ὑπεναντία.

When two quantities are compared, it may be considered *how much* the one is greater than the other, what is their *difference*; or it may be considered *how many times* the one is contained in the other, what is their *quotient*. The former relation of the two quantities is called their *arithmetical ratio*; the latter their *geometrical ratio*.

Let now three magnitudes, lines or numbers, a, b, c, be taken. If $a-b=b-c$, the three magnitudes are in arithmetical proportion; but if $a:b::b:c$, they are in geometrical proportion.[66] In the latter case, it follows at once, from the second theorem of Thales (Euclid, VI., 4), that

$$a-b:b-c::a:b;$$

whereas in the former case we have plainly

$$a-b:b-c::a:a.$$

This might have suggested the consideration of three magnitudes, so taken that

$$a-b:b-c::a:c:$$

three such magnitudes are in harmonical proportion.

The probability of the correctness of this view is indicated by the consideration of the three later proportions—

$a:c::b-c:a-b$ . . . the contrary of the harmonical;

$\left.\begin{array}{l} b:c::b-c:a-b \\ a:b::b-c:a-b \end{array}\right\}$ . . . the contrary of the geometrical.

The discovery of these proportions is attributed to Hippasus, Archytas, and Eudoxus.[67]

[66] In *lines* we may have $c=a-b$, or $a:b:a-b$. This particular case, in which the geometrical and arithmetical ratios both occur in the same proportion, is worth noticing. The line a is then the sum of the other two lines, and is said to be cut in extreme and mean ratio. This section, as we have seen, has arisen out of the construction of the regular pentagon, and we learn from Kepler that it was called by the moderns, on account of its many wonderful properties, *sectio divina, et proportio divina*. He sees in it a fine image of generation, since the addition to the line of its greater part produces a new line cut similarly, and so on. See Kepleri, *Opera Omnia*, ed. Frisch, vol. v., pp. 90 and 187 (*Harmonia Mundi*); also vol. I., p. 377 (*Literae de Rebus Astrologicis*). The pentagram might be taken as the image of all this, as each of its sides and part of a side are cut in this *divine* proportion.

[67] Iambl. in Nic. *Arith.*, pp. 142, 159, 163. See above, p. 4.

We have seen also (*p*) that a knowledge of the so-called most perfect or musical proportion, which comprehends in it all the former ratios, is attributed by Iamblichus to Pythagoras—

$$a : \frac{a+b}{2} :: \frac{2ab}{a+b} : b.$$

We have also seen (*q*) that a knowledge of the doctrine of arithmetical progressions is attributed to Pythagoras. This much at least seems certain, that he was acquainted with the summation of the natural numbers, the odd numbers, and the even numbers, all of which are capable of geometrical representation.

Montucla says that Pythagoras laid the foundation of the doctrine of *Isoperimetry* by proving that of all figures having the same perimeter the circle is the greatest, and that of all solids having the same surface the sphere is the greatest.[68]

There is no evidence to support this assertion, though it is repeated by Chasles, Arneth, and others; it rests merely on an erroneous interpretation of the passage (*s*) in Diogenes Laertius, which says only that "of all solid figures the sphere is the most beautiful; and of all plane figures, the circle." Pythagoras attributes perfection and beauty to the sphere and circle on account of their regularity and uniformity. That this is the true signification of the passage is confirmed by Plato in the Timaeus,[69] when speaking of the Pythagorean cosmogony.[70].

We must also deny to Pythagoras and his school a knowledge of the conic sections, and, in particular, of the quadrature of the parabola, attributed to him by some

[68] "Suivant Diogène, dont le texte est ici fort corrumpu, et probablement transposé, il ébaucha aussi la doctrine des Isopérimètres, en démontrant que de toutes les figures de même contour, parmi les figures planes, c'est le cercle qui est la plus grande, et parmi les solides, la sphère."—Montucla, *Histoire des Mathématiques*, tom. I., p. 113.

[69] *Timaeus*, 33, B., vol. VII., ed. Stallbaum, p. 129.

[70] See Bretschneider, *Geom. vor Eukl.*, pp. 89, 90.

authors, and we have already noticed the misconception which gave rise to this erroneous conclusion.[71]

Let us now see what conclusions can be drawn from the foregoing examination of the mathematical work of Pythagoras and his school, and thus form an estimate of the state of geometry about 480 B.C.

First, then, as to *matter* :—

It forms the bulk of the first two books of Euclid, and includes, further, a sketch of the doctrine of proportion—which was probably limited to commensurable magnitudes—together with some of the contents of the sixth book. It contains, too, the discovery of the irrational (ἄλογον), and the construction of the regular solids; the latter requiring the description of certain regular polygons—the foundation, in fact, of the fourth book of Euclid.

The properties of the circle were not much known at this period, as may be inferred from the fact that not one remarkable theorem on this subject is mentioned; and we shall see later that Hippocrates of Chios did not know the theorem—that the angles in the same segment of a circle are equal to each other. Though this be so, there is, as we have seen, a tradition (*t*) that the problem of the quadrature of the circle also engaged the attention of the Pythagorean school—a problem which they probably derived from the Egyptians.[72]

Secondly, as to *form* :—

The Pythagoreans first severed geometry from the needs of practical life, and treated it as a liberal science, giving

[71] See above, p. 25, note 27.

[72] This problem is considered in the Papyrus Rhind, pp, 97, 98, 117. The point of view from which it was regarded by the Egyptians was different from that of Archimedes. Whilst he made it to depend on the determination of the ratio of the circumference to the diameter, they sought to find from the diameter the side of a square whose area should be equal to that of the circle. Their approximation was as follows :—The diameter being divided into nine equal parts, the side of the equivalent square was taken by them to consist of eight of those parts.

definitions, and introducing the manner of proof which has ever since been in use. Further, they distinguished between *discrete* and *continuous* quantities, and regarded geometry as a branch of mathematics, of which they made the fourfold division that lasted to the Middle Ages—the *quadrivium* (fourfold way to knowledge) of Boetius and the scholastic philosophy. And it may be observed, too, that the name of mathematics, as well as that of philosophy, is ascribed to them.

Thirdly, as to *method*:—

One chief characteristic of the mathematical work of Pythagoras was the combination of arithmetic with geometry. The notions of an equation and a proportion—which are common to both, and contain the first germ of algebra—were, as we have seen, introduced amongst the Greeks by Thales. These notions, especially the latter, were elaborated by Pythagoras and his school, so that they reached the rank of a true scientific method in their Theory of Proportion. To Pythagoras, then, is due the honour of having supplied a method which is common to all branches of mathematics, and in this respect he is fully comparable to Descartes, to whom we owe the decisive combination of algebra with geometry.

It is necessary to dwell on this at some length, as modern writers are in the habit of looking on proportion as a branch of arithmetic[73]—no doubt on account of the arithmetical point of view having finally prevailed in it—whereas for a long period it bore much more the marks of its geometrical origin.[74]

That proportion was not thus regarded by the ancients, merely as a branch of arithmetic, is perfectly plain. We

[73] Bretschneider (*Geom. vor Eukl.*, p. 74) and Hankel (*Gesch. der Math.*, p. 104) do so, although they are treating of the history of Greek geometry, which is clearly a mistake.

[74] On this see A. Comte, *Politique Positive*, vol. III., ch. iv., p. 300.

learn from Proclus that "Eratosthenes looked on proportion as the bond (σύνδεσμον) of mathematics."[75]

We are told, too, in an anonymous scholium on the Elements of Euclid, which Knoche attributes to Proclus, that the fifth book, which treats of proportion, is common to geometry, arithmetic, music, and, in a word, to all mathematical science.[76]

And Kepler, who lived near enough to the ancients to reflect the spirit of their methods, says that one part of geometry is concerned with the comparison of figures and quantities, whence proportion arises ("unde proportio existit"). He also adds that arithmetic and geometry afford mutual aid to each other, and that they cannot be separated.[77]

And since Pythagoras they have never been separated. On the contrary, the union between them, and indeed, between the various branches of mathematics, first instituted by Pythagoras and his school, has ever since become more intimate and profound. We are plainly in presence of not merely a great mathematician, but of a great philosopher. It has been ever so—the greatest steps in the development of mathematics have been made by philosophers.

Modern writers are surprised that Thales, and indeed all the principal Greek philosophers prior to Pythagoras, are named as his masters. They are suprised, too, at the extent of the travels attributed to him. Yet there is no

[75] Proclus, ed. Friedlein, p. 43.

[76] Euclidis *Elem.* Graece ed. ab E. F. August, pars ii., p. 328, Berolini, 1829. Dr. J. H. Knoche, *op. cit.*, p. 10.

[77] "Et quidem geometriae theoreticae initio hujus tractatus duas fecimus partes, unam de magnitudinibus, quatenus fiunt figurae, alteram de comparatione figurarum et quantitatum, unde proportio existit.

"Hae duae scientiae, arithmetica et geometria speculativa, mutuas tradunt operas nec ab invicem separari possunt, quamvis et arithmetica sit principium cognitionis."—Kepleri *Opera Omnia*, ed. Dr. h. Frisch, vol. VIII., p. 160, Francofurti, 1870.

cause to wonder that he was believed by the ancients to have had these philosophers as his teachers, and to have extended his travels so widely in Greece, Egypt, and the East, in search of knowledge, for—like the geometrical figures on whose properties he loved to meditate—his philosophy was many-sided, and had points of contact with all these :—

He introduced the knowledge of arithmetic from the Phoenicians, and the doctrine of proportion from the Babylonians.

Like Moses, he was learned in all the wisdom of the Egyptians, and carried their geometry and philosophy into Greece.

He continued the work commenced by Thales in abstract science, and invested geometry with the form which it has preserved to the present day.

In establishing the existence of the regular solids he showed his deductive power; in investigating the elementary laws of sound he proved his capacity for induction; and in combining arithmetic with geometry, and thereby instituting the theory of proportion, he gave an instance of his philosophic power.

These services, though great, do not form, however, the chief title of this Sage to the gratitude of mankind. He resolved that the knowledge which he had acquired with so great labour, and the doctrine which he had taken such pains to elaborate, should not be lost; and, as a husbandman selects good ground, and is careful to prepare it for the reception of the seed, which he trusts will produce fruit in due season, so Pythagoras devoted himself to the formation of a society *d'élite*, which would be fit for the reception and transmission of his science and philosophy; and thus became one of the chief benefactors of humanity, and earned the gratitude of countless generations.

His disciples proved themselves worthy of their high mission. We have had already occasion to notice their noble self-renunciation, which they inherited from their master.

The moral dignity of these men is further shown by their admirable maxim—a maxim conceived in the spirit of true social philosophers—*a figure and a step; but not a figure and three oboli* (σχᾶμα καὶ βᾶμα, ἀλλ' οὐ σχᾶμα καὶ τριώβολον).[78]

Such, then, were the men by whom the first steps in mathematics—the first steps ever the most difficult—were made.

In the next chapter we shall notice the events which led to the publication, through Hellas, of the results arrived at by this immortal School.

[78] Proclus, ed. Friedlein, p. 84. Taylor's *Commentaries of Proclus*, vol. I., p. 113. Taylor, in a note on this passage, says:—"I do not find this aenigma among the Pythagoric symbols which are extant, so that it is probably nowhere mentioned but in the present work."

Taylor is not correct in this statement. This symbol occurs in Iamblichus. See Iambl., *Adhortatio ad Philosophiam*, ed. Kiessling, *Symb.* XXXVI., cap. xxi., p. 317; also *Expl.* p. 374: προτίμα τὸ σχῆμα καὶ βῆμα τοῦ σχῆμα καὶ τριώβολον.

CHAPTER III.*

THE GEOMETERS OF THE FIFTH CENTURY B.C.—HIPPOCRATES OF CHIOS: DEMOCRITUS.

State of Hellas during the Persian Wars.—Pre-eminence of Athens.—The Eleatic School and the Atomic Philosophy.—Hippocrates of Chios.—Notices of him and of his work.—Was the first writer of the "Elements," and first squared the Lune.—His probable date.—The Quadrature of the Circle and of the Lune.—Extract from the Commentary of Simplicius on the Physics of Aristotle; derived by him partly from Alexander of Aphrodisias and partly from the History of Geometry of Eudemus.—Notices of the work done by other Geometers of this period.—Democritus.—His mathematical writings.—Problem of the Duplication of the Cube.—Reduced by Hippocrates to: the finding of two Mean Proportionals between two Given Straight Lines.—Probable relation of this problem to the Pythagorean Cosmology.—Its influence on the development of Geometry.—The Trisection of an Angle.—The Quadratrix.—Hippias of Elis.—Method of Exhaustions erroneously attributed to Hippocrates.—Probable origin of his discovery concerning the Quadrature of the Lune.—Though the principal Geometer of this period, the judgment of the ancients on him was not altogether favourable.—Suggested explanation of this.

THE first twenty years of the fifth century before the Christian era was a period of deep gloom and despondency throughout the Hellenic world. The Ionians had revolted, and were conquered for the third time; this time, how-

* In the Introduction, note 1, I acknowledged my obligations to the works of Bretschneider and Hankel: I have again made use of them in the preparation of this chapter. Since it was written, I have received from Dr. Moritz Cantor, of Heidelberg, the portion of his *History of Mathematics* which treats of the Greeks (*Vorlesungen über Geschichte der Mathematik*, von Moritz Cantor, Erster Band. Von den ältesten Zeiten bis zum Jahre 1200 n. Chr. Leipzig, 1880 (Teubner)). To the list of new editions of ancient mathematical works given in the note referred to above, I have to add: Theonis Smyrnaei *Expositio rerum Mathematicarum ad legendum Platonem utilium*. Recensuit Eduardus Hiller, Lipsiae, 1878 (Teubner); Pappi Alexandrini *Collectionis quae supersunt*, &c., instruxit F. Hultsch, vol. III., Berolini, 1878 (to the latter the editor has appended an *Index Graecitatis*, a valuable addition; for, as he remarks, "Mathematicam Graecorum dictionem nemo adhuc in lexici formam redegit." Praef., vol. III.,

ever, the conquest was complete and final: they were overcome by sea as well as by land. Miletus, till then the chief city of Hellas, and rival of Tyre and Carthage, was taken and destroyed; the Phœnician fleet ruled the sea, and the islands of the Ægean became subject to Persia. The fall of Ionia, and the maritime supremacy of the Phœnicians, involving the interruption of Greek commerce, must have exercised a disastrous influence on the cities of Magna Graecia.[1] The events which occurred there after the destruction of Sybaris are involved in great obscurity. We are told that some years after this event there was an uprising of the democracy—which had been repressed under the influence of the Pythagoreans—not only in Crotona, but also in the other cities of Magna Graecia. The Pythagoreans were attacked, and their Senate-houses (συνέδρια) were burned; the whole country was thrown into a state of confusion and anarchy; the Pythagorean Brotherhood was suppressed, and the chief men in each city perished.

The Italic Greeks, as well as the Ionians, ceased to prosper.

Towards the end of this period Athens was in the hands of the Persians, and Sicily was threatened by the Carthaginians. Then followed the glorious struggle; the gloom was dispelled, the war which had been at first

tom. II.); Archimedis *Opera omnia cum commentariis Eutocii.* E codice Florentino recensuit, Latine vertit notisque illustravit J. L. Heiberg, Dr. Phil. Vol. I., Lipsiae, 1880 (Teubner). Since the above was in type, the following work has been published: *An Introduction to the Ancient and Modern Geometry of Conics: being a Geometrical Treatise on the Conic Sections, with a collection of Problems and Historical Notes and Prolegomena.* By Charles Taylor, M.A., Fellow of St. John's College, Cambridge. Cambridge, 1881. The matter of the *Prolegomena*, pp. xvii–lxxxviii, is historical.

[1] The names *Ionian Sea*, and *Ionian Isles* still bear testimony to the intercourse between these cities and Ionia. The writer of the article in Smith's *Dictionary of Geography* thinks that the name Ionian Sea was derived from Ionians residing, in very early times, on the west coast of the Peloponnesus. Is it not more probable that it was so called from being the highway of the Ionian ships, just as, now-a-days, in a provincial town we have the *London* road?

defensive became offensive, and the Ægean sea was cleared of Phœnicians and pirates. A solid basis was thus laid for the development of Greek commerce and for the interchange of Greek thought, and a brilliant period followed—one of the most memorable in the history of the world.

Athens now exercised a powerful attraction on all that was eminent in Hellas, and became the centre of the intellectual movement. Anaxagoras settled there, and brought with him the Ionic philosophy, numbering Pericles and Euripides amongst his pupils. Many of the dispersed Pythagoreans no doubt found a refuge in that city, always hospitable to strangers: subsequently the Eleatic philosophy was taught there by Parmenides and Zeno. Eminent teachers flocked from all parts of Hellas to the Athens of Pericles. All were welcome; but the spirit of Athenian life required that there should be no secrets, whether confined to priestly families[2] or to philosophic sects: everything should be made public.

In this city, then, geometry was first published; and with that publication, as we have seen, the name of Hippocrates of Chios is connected.

Before proceeding, however, to give an account of the work of Hippocrates of Chios, and the geometers of the fifth century before the Christian era, we must take a cursory glance at the contemporaneous philosophical movement. Proclus makes no mention of any of the philosophers of the Eleatic School in the summary of the history of geometry which he has handed down—they seem, indeed, not to have made any addition to geometry or astronomy, but rather to have affected a contempt for both these sciences—and most writers[3] on the history of

[2] *E.g.* the Asclepiadae. See Curtius, *History of Greece*, Engl. transl., vol. II. p. 510.

[3] Not so Hankel, whose views as to the influence of the Eleatic philosophy

mathematics either take no notice whatever of that School, or merely refer to it as outside their province. Yet the visit of Parmenides and Zeno to Athens (*circ.* 450 B.C.), the invention of dialectics by Zeno, and his famous polemic against multiplicity and motion, not only exercised an important influence on the development of geometry at that time, but, further, had a lasting effect on its subsequent progress in respect of *method*.[4]

Zeno argued that neither multiplicity nor motion is possible, because these notions lead to contradictory consequences. In order to prove a contradiction in the idea of motion, Zeno argues: "Before a moving body can arrive at its destination it must have arrived at the middle of its path; before getting there it must have accomplished the half of that distance, and so on *ad infinitum*: in short, every body, in order to move from one place to another, must pass through an infinite number of spaces, which is impossible." Similarly he argued that "Achilles cannot overtake the tortoise, if the latter has got any start, because in order to overtake it he would be obliged first to reach every one of the infinitely many places which the tortoise had previously occupied." In like manner, "The flying arrow is always at rest; for it is at each moment only in one place."

Zeno applied a similar argument to show that the notion of multiplicity involves a contradiction. 'If the manifold exists, it must be at the same time infinitely small and infinitely great—the former, because its last

I have adopted. See a fine chapter of his *Gesch. der Math.*, pp. 115 *sq.*, from which much of what follows is taken.

[4] This influence is noticed by Clairaut, *Élémens de Géométrie*, Pref. p. x, Paris, 1741: "Qu' Euclide se donne la peine de démontrer, que deux cercles qui se coupent n'ont pas le même centre, qu'un triangle renfermé dans un autre a la somme de ses côtés plus petite que celle des côtés du triangle dans lequel il est renfermé; on n'en sera pas surpris. Ce Géométre avoit à convaincre des Sophistes obstinés, qui se faisoient gloire de se refuser aux vérités les plus évidentes: il falloit donc qu'alors la Géométrie eût, comme la Logique, le secours des raisonnemens en forme, pour fermer la bouche à la chicanne."

divisions are without magnitude; the latter, on account of the infinite number of these divisions.' Zeno seems to have been unable to see that if $xy = a$, x and y may both vary, and that the number of parts taken may make up for their minuteness.

Subsequently the Atomists endeavoured to reconcile the notions of unity and multiplicity; stability and motion; permanence and change; being and becoming—in short, the Eleatic and Ionic philosophy. The atomic philosophy was founded by Leucippus and Democritus; and we are told by Diogenes Laertius that Leucippus was a pupil of Zeno: the filiation of this philosophy to the Eleatic can, however, be seen independently of this statement. In accordance with the atomic philosophy magnitudes were considered to be composed of indivisible elements (ἀτόμοι) in finite numbers; and indeed Aristotle—who, a century later, wrote a treatise on *Indivisible Lines* (περὶ ἀτόμων γραμμῶν), in order to show their mathematical and logical impossibility—tell us that Zeno's disputation was taken as compelling such a view.[5] We shall see, too, that, in Antiphon's attempt to square the circle, it is assumed that straight and curved lines are ultimately reducible to the same indivisible elements.[6]

Insuperable difficulties were found, however, in this conception; for no matter how far we proceed with the division, the distinction between the straight and curved still exists. A like difficulty had been already met with in the case of straight lines themselves, for the incommensurability of certain lines had been established by the Pythagoreans. The diagonal of a square, for example, cannot be made up of submultiples of the side, no matter how minute these submultiples may be. It is possible that Democritus may have attempted to get over this diffi-

[5] Arist., *De insecab. lineis*, p. 968[a], ed. Bekker.

[6] Vid. Bretsch., *Geom. vor Eukl.*, p. 101, *et infra*, p. 66.

culty, and reconcile incommensurability with his atomic theory; for we are told by Diogenes Laertius that he wrote on incommensurable lines and solids (περὶ ἀλόγων γραμμῶν καὶ ναστῶν).[7]

The early Greek mathematicians, troubled no doubt by these paradoxes of Zeno, and finding the progress of mathematics impeded by their being made a subject of dialectics, seem to have avoided all these difficulties by banishing from their science the idea of the Infinite—the infinitely small as well as the infinitely great (*vid.* Euclid, Book V., Def. 4). They laid down as axioms that any quantity may be divided *ad libitum*; and that, if two spaces are unequal, it is possible to add their difference to itself so often that every finite space can be surpassed.[8] According to this view, there can be no infinitely small difference which being multiplied would never exceed a finite space.

HIPPOCRATES of Chios, who must be distinguished from his contemporary and namesake, the great physician of Cos, was originally a merchant. All that we know of him is contained in the following brief notices:—

(*a*). Plutarch tell us that Thales, and Hippocrates the mathematician, are said to have applied themselves to commerce.[9]

(*b*). Aristotle reports of him: "It is well known that persons, stupid in one respect, are by no means so in others (there is nothing strange in this: so Hippocrates, though skilled in geometry, appears to have been in other respects weak and stupid; and he lost, as they say, through his simplicity, a large sum of money by the fraud of the collectors of customs at Byzantium (ὑπὸ τῶν ἐν Βυζαντίῳ πεντηκοστολόγων))."[10]

7 Diog. Laert., IX. vii., 47, ed. Cobet, p. 239.

8 Archim., *de quadr. parab.*, p. 18, ed. Torelli.

9 *Vit. Solonis*, II.

10 Arist., *Eth. ad Eud.*, VII., c. xiv., p. 1247^a, 15, ed. Bekker.

(*c*). Johannes Philoponus, on the other hand, relates that Hippocrates of Chios, a merchant, having fallen in with a pirate vessel, and having lost everything, went to Athens to prosecute the pirates, and staying there a long time on account of the prosecution, frequented the schools of the philosophers, and arrived at such a degree of skill in geometry, that he endeavoured to find the quadrature of the circle.[11]

(*d*). We learn from Eudemus that Œnopides of Chios was somewhat junior to Anaxagoras, and that after these Hippocrates of Chios, who first found the quadrature of the lune, and Theodorus of Cyrene, became famous in geometry; and that Hippocrates was the first writer of Elements.[12]

(*e*). He also taught, for Aristotle says that his pupils, and those of his disciple Æschylus, expressed themselves concerning comets in a similar way to the Pythagoreans.[13]

(*f*). He is also mentioned by Iamblichus, along with Theodorus of Cyrene, as having divulged the geometrical *arcana* of the Pythagoreans, and thereby having caused mathematics to advance (ἐπέδωκε δὲ τὰ μαθήματα, ἐπεὶ ἐξενηνέχθησαν δισσοὶ προαγόντε, μάλιστα Θεόδωρός τε ὁ Κυρηναῖος, καὶ Ἱπποκράτης ὁ Χῖος).[14]

(*g*). Iamblichus goes on to say that the Pythagoreans allege that geometry was made public thus: one of the Pythagoreans lost his property; and he was, on account of his misfortune, allowed to make money by teaching geometry.[15]

(*h*). Proclus, in a passage quoted *supra* (p. 41, note 62), ascribes to Hippocrates the method of reduction (ἀπαγωγή). Proclus defines ἀπαγωγή to be "a transi-

[11] Philoponus, *Comm. in Arist. phys. ausc.*, f. 13. Brand., *Schol. in Arist.*, p. 327^b, 44.

[12] Proclus, ed. Friedlein, p. 66.

[13] Arist., *Meteor.*, I., vi., p. 342^b, 35, ed. Bekker.

[14] Iambl. *de philos. Pythag.* lib. III.; Villoison, *Anecdota Graeca*, II., p. 216.

[15] *Ibid.*; also Iambl., *Vit. Pyth.*, Cap. XVIII., 89.

tion from one problem or theorem to another, which being known or determined, the thing proposed is also plain. For example: when the duplication of the cube is investigated, geometers reduce the question to another to which this is consequent, *i. e.*, the finding of two mean proportionals, and afterwards they inquire how between two given straight lines two mean proportionals may be found. But Hippocrates of Chios is reported to have been the first inventor of geometrical reduction (ἀπαγωγή); who also squared the lune, and made many other discoveries in geometry, and who was excelled by no other geometer in his powers of construction."[16]

(*i*). Eratosthenes, too, in his letter to King Ptolemy III., Euergetes, which has been handed down to us by Eutocius, after relating the legendary origin of the celebrated problem of the duplication of the cube, tells us that after geometers had for a long time been quite at a loss how to solve the question, it first occurred to Hippocrates of Chios that if between two given lines, of which the greater is twice the less, he could find two mean proportionals, then the problem of the duplication of the cube would be solved. But thus, Eratosthenes adds, the problem is reduced to another which is no less difficult.[17]

(*k*). Eutocius, in his commentary on Archimedes (*Circ. Dimens.* Prop. 1), tells us that Archimedes "wished to show that a circle is equal to a certain rectilineal area, a thing which had been of old investigated by illustrious philosophers.[18] For it is evident that this is the problem concerning which Hippocrates of Chios and Antiphon, who carefully searched after it, invented the false reasonings which, I think, are well known to those who have looked into the *History of Geometry* of Eudemus and the *Keria* (Κηρίων) of Aristotle."[19]

16 Proclus, ed. Friedlein, pp. 212, 213.
17 Archim., ex recens. Torelli, p. 144: Oxon. 1792.
18 Anaxagoras, for example.
19 Archim., ex recens. Torelli, p. 204.

On the passage (*f*) quoted above, from Iamblichus, is based the statement of Montucla, which has been repeated since by recent writers on the history of mathematics,[20] that Hippocrates was expelled from a school of Pythagoreans for having taught geometry for money.[21]

There is no evidence whatever for this statement, which is, indeed, inconsistent with the passage (*g*) of Iamblichus which follows. Further, it is even possible that the person alluded to in (*g*) as having been allowed to make money by teaching geometry may have been Hippocrates himself; for—

1. He learned from the Pythagoreans;
2. He lost his property through misfortune;
3. He made geometry public, not only by teaching, but also by being the first writer of the Elements.

This misapprehension originated, I think, with Fabricius, who says: "De Hippaso Metapontino adscribam adhuc locum Iamblichi è libro tertio de Philosophia Pythagorica Graece necdum edito, p. 64, ex versione Nic. Scutellii: *Hippasus* (videtur legendum Hipparchus) *ejicitur è Pythagorae schola eo, quod primus sphaeram duodecim angulorum* (Dodecaedron) *edidisset* (adeoque arcanum hoc evulgasset), *Theodorus etiam Cyrenaeus et Hippocrates Chius Geometra ejiciuntur qui ex geometria quaestum factitabant.* Confer Vit. Pyth., cap. 34 & 35."[22]

In this passage Fabricius, who, however, had access to

[20] Bretsch., *Geom. vor Eukl.*, p. 93; Hoefer, *Histoire des Math.*, p. 135. Since the above was written, this statement has been reiterated by Cantor, *Gesch. der Math.*, p. 172; and by C. Taylor, *Geometry of Conics*, *Prolegomena*, p. xxviii.

[21] Montucla, *Histoire des Math.*, tom. I., p. 144, 1re ed. 1758; tom. I., p. 152, nouv. ed. an vii.; the statement is repeated in p. 155 of this edition, and Simplicius is given as the authority for it. Iamblichus is, however, referred to by later writers as the authority for it.

[22] Jo. Alberti Fabricii *Bibliotheca Graeca*, ed. Harles, I. p. 848, Hamburgi, 1790.

a manuscript only, falls into several mistakes, as will be seen by comparing it with the original, which I give here :—

Περὶ δ' Ἱππάσου λέγουσιν, ὡς ἦν μὲν τῶν Πυθαγορείων, διὰ δὲ τὸ ἐξενεγκεῖν, καὶ γράψασθαι πρῶτος σφαῖραν, τὴν ἐκ τῶν δώδεκα ἑξαγώνων [πενταγώνων], ἀπόλοιτο κατὰ θάλατταν, ὡς ἀσεβήσας, δόξαν δὲ λάβοι, ὡς εἶναι δὲ πάντα ἐκείνου τοῦ ἀνδρός· προσαγορεύουσι γὰρ οὕτω τὸν Πυθαγόραν, καὶ οὐ καλοῦσιν ὀνόματι. ἐπέδωκε δὲ τὰ μαθήματα, ἐπεὶ ἐξενηνέχθησαν δισσοὶ προαγόντε, μάλιστα Θεόδωρός τε ὁ Κυρηναῖος, καὶ Ἱπποκράτης ὁ Χῖος. λέγουσι δὲ οἱ Πυθαγόρειοι ἐξενηνέχθαι γεωμετρίαν οὕτως· ἀποβαλεῖν τινα τὴν οὐσίαν τῶν Πυθαγορείων· ὡς δὲ τοῦτ' ἠτύχησε, δοθῆναι αὐτῷ χρηματίσασθαι ἀπὸ γεωμετρίας· ἐκαλεῖτο δέ ἡ γεωμετρία πρὸς Πυθαγόρου ἱστορία.[23]

Observe that Fabricius, mistaking the sense, says that Hippasus, too, was expelled. Hippocrates may have been expelled by a School of Pythagoreans with whom he had been associated; but, if so, it was not for teaching geometry for money, but for taking to himself the credit of Pythagorean discoveries—a thing of which we have seen the Pythagoreans were most jealous, and which they even looked on as impious (ἀσεβήσας).[24]

As Anaxagoras was born 499 B.C., and as Plato, after the death of Socrates 399 B.C., went to Cyrene to hear Theodorus (*d*), the lifetime of Hippocrates falls within the fifth century before Christ. As, moreover, there could not have been much commerce in the Ægean during the first quarter of the fifth century, and, further, as the statements of Aristotle and Philoponus [(*b*) and (*c*)] fall in better with the state of affairs during the Athenian supremacy—even though we do not accept the suggestion of Bretschneider,

[23] Iambl., *de philos. Pyth.*, lib. III.; Villoison, *Anecdota Graeca*, II., p. 216. With the exception of the sentence concerning Hippocrates, the passage, with some modifications, occurs also in Iambl., *Vit. Pyth.*, Cap. XVIII., 88, 89.

[24] See p. 43.

made with the view of reconciling these inconsistent statements, that the ship of Hippocrates was taken by Athenian pirates[25] during the Samian war (440 B.C.), in which Byzantium took part—we may conclude with certainty that Hippocrates did not take up geometry until after 450 B.C. We have good reason to believe that at that time there were Pythagoreans settled at Athens. Hippocrates, then, was probably somewhat senior to Socrates, who was a contemporary of Philolaus and Democritus.

The paralogisms of Hippocrates, Antiphon, and Bryson, in their attempts to square the circle, are referred to and contrasted with one another in several passages of Aristotle[26] and of his commentators—Themistius,[27] Jo. Philoponus,[28] and Simplicius. Simplicius has preserved in his *Commentary on the Physics of Aristotle* a pretty full and partly literal extract from the *History of Geometry* of Eudemus, which contains an account of the work of Hippocrates and others in relation to this problem. The greater part of this extract had been almost entirely overlooked by writers on the history of Mathematics, until Bretschneider[29] republished the Greek text, having carefully revised and emended it. He also supplied the necessary diagrams, some of which were wanting, and added explanatory and critical notes. This extract is interesting and important, and Bretschneider is entitled to much credit for the pains he has taken to make it intelligible and better known.

[25] Bretsch., *Geom. vor Eukl.*, p. 98.

[26] *De Sophist. Elench.*, 11, pp. 171^b, and 172, ed. Bekker; *Phys. Ausc.*, I., ii., p. 185^a, 14, ed. Bekker.

[27] Themist., f. 16, *Schol.* in Arist., Brand., p. 327^b, 33. *Ibid.*, f. 5, *Schol.*, p. 211^b, 19.

[28] Jo. Philop. f. 25b, *Schol.*, Brand. p. 211^b, 30. *Ibid.*, f. 118, *Schol.*, p. 211^b, 41. *Ibid.*, f. 26b, *Schol.*, p. 212^a, 16.

[29] Bretsch., *Geom. vor Eukl.*, pp. 100–121. [Simplicii *in Aristotelis Physicorum libros quattuor priores Commentaria*, ed. Hermannus Diels, pp. 54–69.]

It is much to be regretted, however, that Simplicius did not merely transmit *verbatim* what Eudemus related, and thus faithfully preserve this oldest fragment of Greek geometry, but added demonstrations of his own, giving references to the Elements of Euclid, who lived a century and a-half later than Hippocrates. Simplicius says: "I shall now put down literally what Eudemus relates, adding only a short explanation by referring to Euclid's Elements, on account of the summary manner of Eudemus, who, according to archaic custom, gives only concise proofs."[30] And in another place he tells us that Eudemus passed over the squaring of a certain lune as evident—indeed, Eudemus was right in doing so—and supplies a lengthy demonstration himself.[31]

Bretschneider and Hankel, overlooking these passages, and disregarding the frequent references to the Elements of Euclid which occur in this extract, have drawn conclusions as to the state of geometry at the time of Hippocrates which, in my judgment, cannot be sustained. Bretschneider notices the great circumstantiality of the construction, and the long-windedness and the over-elaboration of the proofs.[32] Hankel expresses surprise at the fact that this oldest fragment of Greek geometry—150 years older than Euclid's Elements—already bears that character, typically fixed by the latter, which is so peculiar to the geometry of the Greeks.[33]

Fancy a naturalist finding a fragment of the skeleton of some animal which had become extinct, but of which there were living representatives in a higher state of development; and fancy him improving the portion of the skeleton in his hands by making additions to it, so

[30] Bretsch., *Geom. vor Eukl.*, p. 109.

[31] *Ibid.*, p. 113.

[32] *Ibid.*, pp. 130, 131.

[33] Hankel, *Gesch. der Math.*, p. 112.

that it might be more like the skeleton of the living animal; then fancy other naturalists examining the improved fragment with so little attention as to exclaim: "Dear me! how strange it is that the two should be so perfectly alike!"

There is, moreover, much clumsiness, and a want of perspicuity, in the arrangement of the demonstrations—the construction not being clearly stated, but being mixed up with the proof: the proofs, too, which in several instances are plainly supplied by Simplicius—inasmuch as propositions of Euclid's Elements are quoted—are unskilful and wearisome on account of the laboured demonstrations of evident theorems, which are repeated several times under different forms: while, on the other hand, some statements and constructions which stand more in need of explanation are passed over without remark. The conclusion is thus forced on us that Simplicius was but a poor geometer; and we have greater reason, therefore, to regret that he was not content with transmitting the work of Eudemus unaltered.

I shall attempt now to restore this fragment by removing from it everything that seems to me not to be the work of Eudemus, and all reference to Euclid's Elements; and by stating briefly, but at the same time clearly and in order, the several steps of each demonstration. I shall also notice the theorems which are made use of, and the problems whose solution is assumed in it:—

"The difference between false conclusions that can be proved to be *such*, and others which cannot, he [Aristotle] shows by some false reasonings in geometry.[34] Amongst the many persons who have sought the squaring of the circle (that is, to find a square which shall be equal to a circle), both Antiphon and Hippocrates believed that they

[34] ψευδογράφημα, literally a misdelineation, a false reasoning founded on a faulty diagram.

had found it, and were equally mistaken. Antiphon's mistake, on account of his not having started from geometrical principles, as we shall see, cannot be disproved geometrically. That of Hippocrates, on the other hand, since he was deceived although he clung to geometrical principles, can be disproved geometrically. For we must analyse and examine those reasonings only which, preserving the acknowledged principles of the science, lead thus to further conclusions; but there is no use in examining those in which these principles are set aside.'

'Antiphon, having drawn a circle, inscribed in it one of those polygons[35] that can be inscribed: let it be a square. Then he bisected each side of this square, and

through the points of section drew straight lines at right angles to them, producing them to meet the circumference; these lines evidently bisect the corresponding segments of the circle. He then joined the new points of section to the ends of the sides of the square, so that four triangles were formed, and the whole inscribed figure became an octagon. And again, in the same way, he bisected each of the sides of the octagon, and drew from the points of bisection

[35] In Greek mathematical writers, τετράγωνον, as far as I know, always means a square. In this oldest geometrical writing, ἑξάγωνον, ὀκτάγωνον, and πολύγωνον denote *regular* hexagon, octagon, and polygon. This is not the case in the Elements of Euclid, who writes, *e. g.*, πεντάγωνον ἰσόπλευρόν τε καὶ ἰσογώνιον, &c. In Pappus, however, these words, though sometimes used generally, for the most part denote regular figures. The Greeks could do this, for they had the words τετράπλευρον, πεντάπλευρον, &c., for quadrilateral, pentagon, &c.

perpendiculars; he then joined the points where these perpendiculars met the circumference with the extremities of the octagon, and thus formed an inscribed figure of sixteen sides. Again, in the same manner, bisecting the sides of the inscribed polygon of sixteen sides, and drawing straight lines, he formed a polygon of twice as many sides; and doing the same again and again, until he had exhausted the surface, he concluded that in this manner a polygon would be inscribed in the circle, the sides of which, on account of their minuteness, would coincide with the circumference of the circle. But we can substitute for each polygon a square of equal surface; therefore we can, since the surface coincides with the circle, construct a square equal to a circle.'

On this Simplicius observes: 'the conclusion here is manifestly contrary to geometrical principles, not, as Alexander maintains, because the geometer supposes as a principle that a circle can touch a straight line in one point only, and Antiphon sets this aside; for the geometer does not suppose this, but proves it. It would be better to say that it is a principle that a straight line cannot coincide with a circumference, for one without meets the circle in one point only, one within in two points, and not more, and the meeting takes place in single points. Yet, by continually bisecting the space between the chord and the arc, it will never be exhausted, nor shall we ever reach the circumference of the circle, even though the cutting should be continued *ad infinitum*: if we did, a geometrical principle would be set aside, which lays down that magnitudes are divisible *ad infinitum*. And Eudemus, too, says that this principle has been set aside by Antiphon.[36]

'But the squaring of the circle by means of segments, he [Aristotle[37]] says, may be disproved geometrically; he

[36] But Eudemus was a pupil of Aristotle, and Antiphon was a contemporary of Democritus.

[37] *Phys. Ausc.* I., ii., p. 185^a, 16, ed. Bekker.

would rather call the squaring by means of lunes, which Hippocrates found out, one by segments, inasmuch as the lune is a segment of the circle. The demonstration is as follows:—

'Let a semicircle $\alpha\beta\gamma$ be described on the straight line $\alpha\beta$; bisect $\alpha\beta$ in δ; from the point δ draw a perpendicular $\delta\gamma$ to $\alpha\beta$, and join $\alpha\gamma$; this will be the side of the square inscribed in the circle of which $\alpha\beta\gamma$ is the semicircle. On $\alpha\gamma$ describe the semicircle $\alpha\epsilon\gamma$. Now since the square on $\alpha\beta$ is equal to double the square on $\alpha\gamma$ (and since the squares on the diameters are to each other as the respective circles

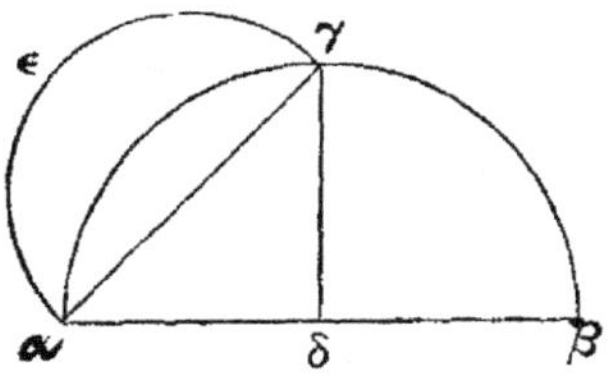

or semicircles), the semicircle $\alpha\gamma\beta$ is double the semicircle $\alpha\epsilon\gamma$. The quadrant $\alpha\gamma\delta$ is, therefore, equal to the semicircle $\alpha\epsilon\gamma$. Take away the common segment lying between the circumference $\alpha\gamma$ and the side of the square; then the remaining lune $\alpha\epsilon\gamma$ will be equal to the triangle $\alpha\gamma\delta$; but this triangle is equal to a square. Having thus shown that the lune can be squared, Hippocrates next tries, by means of the preceding demonstration, to square the circle thus:—

'Let there be a straight line $\alpha\beta$, and let a semicircle be described on it; take $\gamma\delta$ double of $\alpha\beta$, and on it also describe a semicircle; and let the sides of a hexagon, $\gamma\epsilon$, $\epsilon\zeta$, and $\zeta\delta$ be inscribed in it. On these sides describe the semicircles $\gamma\eta\epsilon$, $\epsilon\theta\zeta$, $\zeta\kappa\delta$. Then each of these semicircles described on the sides of the hexagon is equal to the semicircle on $\alpha\beta$, for $\alpha\beta$ is equal to each side of the hexagon. The four semicircles are equal to each other, and together are then four times the semicircle on $\alpha\beta$. But the semicircle on $\gamma\delta$ is also four times that on $\alpha\beta$. The semicircle on $\gamma\delta$

is, therefore, equal to the four semicircles—that on $\alpha\beta$, together with the three semicircles on the sides of the hexagon. Take away from the semicircles on the sides of the hexagon, and from that on $\gamma\delta$, the common segments contained by the sides of the hexagon and the periphery of the semicircle $\gamma\delta$; the remaining lunes $\gamma\eta\epsilon$, $\epsilon\theta\zeta$, and $\zeta\kappa\delta$, together with the semicircle on $\alpha\beta$, will be equal to the trapezium $\gamma\epsilon$, $\epsilon\zeta$, $\zeta\delta$. If we now take away from the trapezium the excess, that is a surface equal to the lunes (for it has been shown that there exists a rectilineal figure

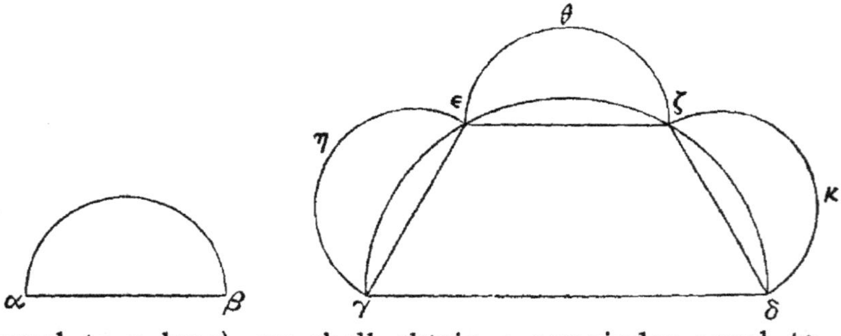

equal to a lune), we shall obtain a remainder equal to the semicircle $\alpha\beta$; we double this rectilineal figure which remains, and construct a square equal to it. That square will be equal to the circle of which $\alpha\beta$ is the diameter, and thus the circle has been squared.

'The treatment of the problem is indeed ingenious; but the wrong conclusion arises from assuming that as demonstrated generally which is not so; for not every lune has been shown to be squared, but only that which stands over the side of the square inscribed in the circle; but the lunes in question stand over the sides of the inscribed hexagon. The above proof, therefore, which pretends to have squared the circle by means of lunes, is defective, and not conclusive, on account of the false-drawn figure (ψευδογράφημα) which occurs in it.[38]

[38] I attribute the above observation on the proof to Eudemus. What follows in Simplicius seems to me not to be his. I have, therefore, omitted the remainder of § 83, and §§ 84, 85, pp. 105–109, Bretsch., *Geom. vor Eukl.*

'Eudemus,[39] however, tells us in his *History of Geometry*, that Hippocrates demonstrated the quadrature of the lune, not merely the lune on the side of the square, but generally, if one might say so: if, namely, the exterior arc of the lune be equal to a semicircle, or greater or less than it. I shall now put down literally (κατὰ λέξιν)[40] what Eudemus relates, adding only a short explanation by referring to Euclid's Elements, on account of the summary manner of Eudemus, who, according to archaic custom, gives concise proofs.

'In the second book of his *History of Geometry*, Eudemus says:—

"The squaring of lunes seeming to relate to an uncommon class of figures was, on account of their relation to the circle, first treated of by Hippocrates, and was rightly viewed in that connection. We may, therefore, more fully touch upon and discuss them. He started with and laid down as the first thing useful for them, that similar segments of circles have the same ratio as the squares on their bases. This he proved by showing that circles have the same ratio as the squares on their diameters. Now, as circles are to each other, so are also similar segments; but similar segments are those which contain the same part of their respective circles, as a semicircle to a semicircle, the third part of a circle to the third part of another circle.[41] For which reason, also, similar segments contain equal angles. The latter are in all semicircles right, in larger segments less than right angles, and so much less as the segments are larger than semicircles; and in smaller segments they are larger than

39 Bretsch., *Geom. vor. Eukl.*, p. 109.

40 Simplicius did not adhere to his intention, or else some transcriber has added to the text.

41 Here τμῆμα seems to be used for sector: indeed, we have seen above that a lune was also called τμῆμα. The word τομεύς, sector, may have been of later origin. The poverty of the Greek language in respect of geometrical terms has been frequently noticed. For example, they had no word for radius, and instead

right angles, and so much larger as the segments are smaller than semicircles. Having first shown this, he described a lune which had a semicircle for boundary, by circumscribing a semicircle about a right-angled isosceles triangle, and describing on the hypotenuse a segment of a circle similar to those cut off by the sides. The segment over the hypotenuse then being equal to the sum of those on the two other sides, if the common part of the triangle which lies over the segment on the base be added

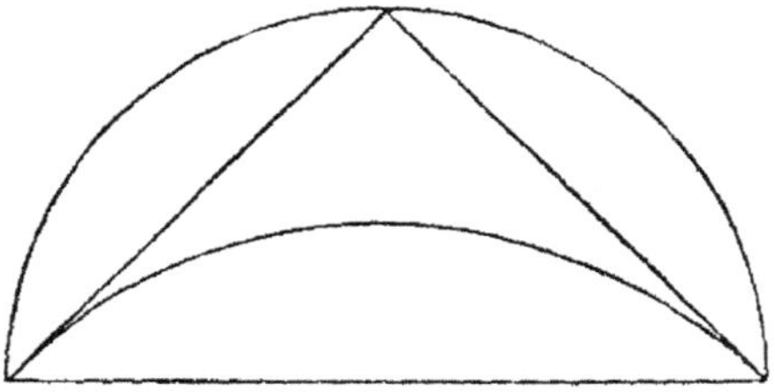

to both, the lune will be equal to the triangle. Since the lune, then, has been shown to be equal to a triangle, it can be squared. Thus, then, Hippocrates, by taking for the exterior arc of the lune that of a semicircle, readily squares the lune.

"Hippocrates next proceeds to square a lune whose exterior arc is greater than a semicircle. In order to do so, he constructs a trapezium[42] having three sides equal to each other, and the fourth—the greater of the two parallel sides—such that the square on it is equal to three times

used the periphrasis ἡ ἐκ τοῦ κέντρου. Again, Archimedes nowhere uses the word parabola; and as to the imperfect terminology of the geometers of this period, we have the direct statement of Aristotle, who says: καὶ τὸ ἀνάλογον ὅτι ἐναλλάξ, ᾗ ἀριθμοὶ καὶ ᾗ γραμμαὶ καὶ ᾗ στερεὰ καὶ ᾗ χρόνοι, ὥσπερ ἐδείκνυτό ποτε χωρίς, ἐνδεχόμενόν γε κατὰ πάντων μιᾷ ἀποδείξει δειχθῆναι· ἀλλὰ διὰ τὸ μὴ εἶναι ὠνομασμένον τι πάντα ταῦτα ἕν, ἀριθμοί μήκη χρόνος στερεά, καὶ εἴδει διαφέρειν ἀλλήλων, χωρὶς ἐλαμβάνετο. νῦν δὲ καθόλου δείκνυται· οὐ γὰρ ᾗ γραμμαὶ ἢ ᾗ ἀριθμοὶ ὑπῆρχεν, ἀλλ' ᾗ τοδί, ὃ καθόλου ὑποτίθενται ὑπάρχειν.—Aristot., *Anal., post.*, I., v., p. 74ª, 17, ed. Bekker. This passage is interesting in another respect also, as it contains the germ of Algebra.

[42] Trapezia, like this, cut off from an isosceles triangle by a line parallel to the base, occur in the Papyrus Rhind.

that on any other side; he circumscribes a circle about the trapezium, and on its greatest side describes a segment of a circle similar to those cut off from the circle by the three equal sides.[43] By drawing a diagonal of the trapezium, it will be manifest that the section in question is greater than a semicircle, for the square on this straight line subtending two equal sides of the trapezium must be greater

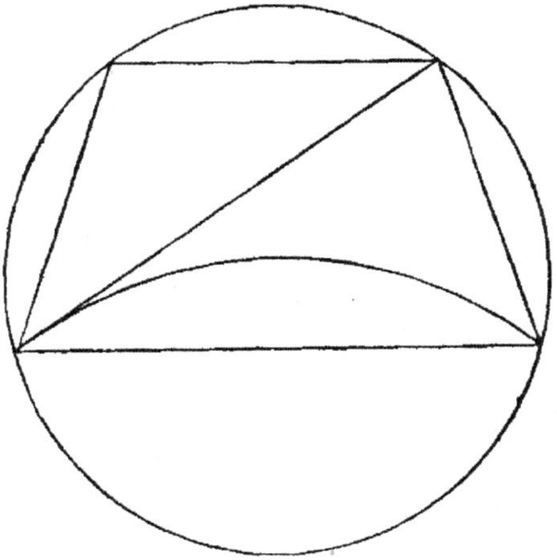

than twice the square on either of them, or than double the square on the third equal side: the square on the greatest side of the trapezium, which is equal to three times the square on any one of the other sides, is therefore less than the square on the diagonal and the square on the third equal side. Consequently, the angle subtended by the greatest side of the trapezium is acute, and the segment which contains it is, therefore, greater than a semicircle: but this is the exterior boundary of the lune. Simplicius tells us that Eudemus passed over the squaring

[43] Then follows a proof, which I have omitted, that the circle can be circumscribed about the trapezium. This proof is obviously supplied by Simplicius, as is indicated by the change of person from ὑποτίθεται to δείξεις, as well as by the reference to Euclid, I. 9. A few lines lower there is a gap in the text, as Bretschneider has observed; but the gap occurs in the work of Simplicius, and not of Eudemus as Bretschneider has erroneously supposed.—*Geom. vor Eukl.*, p. 111, and note.

of this lune, he supposes, because it was evident, and he supplies it himself.[44]

"Further, Hippocrates shows that a lune with an exterior arc less than a semicircle can be squared, and gives the following construction for the description of such a lune:[45]—

"Let $\alpha\beta$ be the diameter of a circle whose centre is κ; let $\gamma\delta$ cut $\beta\kappa$ in the point of bisection γ, and at right angles; through β draw the straight line $\beta\zeta\epsilon$, so that the part of it, $\zeta\epsilon$, intercepted between the line $\gamma\delta$ and the circle shall be such that two squares on it shall be equal to three squares on the radius $\beta\kappa$;[46] join $\kappa\zeta$, and produce it to meet the straight line drawn through ϵ parallel to $\beta\kappa$, and let them meet at η; join $\kappa\epsilon$, $\beta\eta$ (these lines will be equal); describe then a circle round the trapezium $\beta\kappa\epsilon\eta$; also, circumscribe a circle about the triangle $\epsilon\zeta\eta$. Let the centres of these circles be λ and μ respectively.

"Now, the segments of the latter circle on $\epsilon\zeta$ and $\zeta\eta$ are similar to each other, and to each of the segments of the

[44] Bretsch., *Geom. vor Eukl.*, p. 113, § 88. I have omitted it, as not being the work of Eudemus.

[45] The whole construction, as Bretschneider has remarked, is quite obscure and defective. The main point on which the construction turns is the determination of the straight line $\beta\zeta\epsilon$, and this is nowhere given in the text. The determination of this line, however, can be inferred from the statement in p. 114, *Geom. vor Eukl.*, that 'it is assumed that the line $\epsilon\zeta$ inclines towards β'; and the further statement, in p. 117, that 'it is assumed that the square on $\epsilon\zeta$ is once and a-half the square on the radius.' In order to make the investigation intelligible, I have commenced by stating how this line $\beta\zeta\epsilon$ is to be drawn. I have, as usual, omitted the proofs of Simplicius.

Bretschneider, p. 114, notices the archaic manner in which lines and points are denoted in this investigation—ἡ [εὐθεῖα] ἐφ' ᾗ AB, τὸ [σημεῖον] ἐφ' οὗ K—and infers from it that Eudemus is quoting the very words of Hippocrates. I have found this observation useful in aiding me to separate the additions of Simplicius from the work of Eudemus. The inference of Bretschneider, however, cannot I think be sustained, for the same manner of expression is to be found in Aristotle.

[46] The length of the line $\epsilon\zeta$ can be determined by means of the theorem of Pythagoras (Euclid, I. 47), coupled with the theorem of Thales (Euclid, III., 31). Then, produce the line $\epsilon\zeta$ thus determined so that the rectangle under the whole line thus produced and the part produced shall be equal to the square on the radius; or, in archaic language, apply to the line $\epsilon\zeta$ a rectangle which shall be equal to the square on the radius, and which shall be *excessive* by a square—a Pythagorean problem, as Eudemus tells us. (See pp. 24, 41). If the calculation

former circle on the equal straight lines $\epsilon\kappa$, $\kappa\beta$, $\beta\eta$;[47] and, since twice the square on $\epsilon\zeta$ is equal to three times the square on $\kappa\beta$, the sum of the two segments on $\epsilon\zeta$, and $\zeta\eta$ is

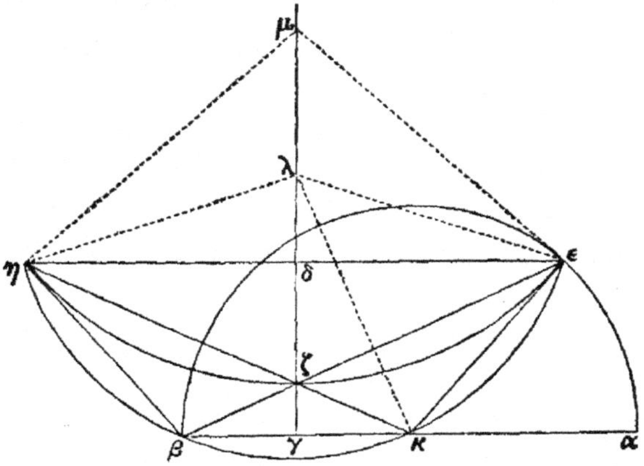

equal to the sum of the three segments on $\epsilon\kappa$, $\kappa\beta$, $\beta\eta$; to each of these equals add the figure bounded by the straight lines $\epsilon\kappa$, $\kappa\beta$, $\beta\eta$, and the arc $\eta\zeta\epsilon$, and we shall have the lune whose exterior arc is $\epsilon\kappa\beta\eta$ equal to the

be made by this method, or by the solution of a quadratic equation, we find

$$\beta\epsilon = \frac{\beta\kappa}{2}\left(\sqrt{\frac{3}{2}}+\sqrt{\frac{11}{2}}\right).$$

Bretschneider makes some slip, and gives

$$\epsilon\beta = \frac{\beta\kappa}{2}\left(\sqrt{\frac{11}{3}}-1\right).$$

Geom. vor Eukl., p. 115, note.

47 Draw lines from the points ϵ, κ, β, and η to λ, the centre of the circle described about the trapezium; and from ϵ and η to μ, the centre of the circle circumscribed about the triangle $\epsilon\zeta\eta$; it will be easy to see, then, that the angles subtended by $\epsilon\kappa$, $\kappa\beta$, $\eta\beta$ at λ are equal to each other, and to each of the angles subtended by $\epsilon\zeta$ and $\zeta\eta$ at μ. The similarity of the segments is then inferred; but observe, that in order to bring this under the definition of similar segments given above, the word *segment* must be used in a large signification; and that further, it requires rather the converse of the definition, and thus raises the difficulty of incommensurability.

The similarity of the segments might also be inferred from the equality of the alternate angles ($\epsilon\eta\zeta$ and $\eta\kappa\beta$, for example). In p. 47, I stated, following Bretschneider and Hankel, that Hippocrates of Chios did not know the theorem that the angles in the same segment of a circle are equal. But if the latter method of proving the similarity of the segments in the construction to which the present note refers was that used by Hippocrates, the statement in question would have to be retracted.

rectilineal figure composed of the three triangles ζβη, ζβκ, ζκε.[48]

"That the exterior arc of this lune is smaller than a semicircle, Hippocrates proves, by showing that the angle εκη lying within the exterior arc of the segment is obtuse, which he does thus: Since the square on εζ is once and a-half the square on the radius βκ or κε, and since, on account of the similarity of the triangles βκε and βζκ, the square on κε is greater than twice the square on κζ,[49] it follows that the square on εζ is greater than the squares on εκ and κζ together. The angle εκη is therefore obtuse, and consequently the segment in which it lies is less than a semicircle.

"Lastly, Hippocrates squared a lune and a circle together, thus: Let two circles be described about the centre κ, and let the square on the diameter of the exterior be six times that of the interior. Inscribe a hexagon αβγδεζ in the inner circle, and draw the radii κα, κβ, κγ, and produce them to the periphery of the exterior circle; let them meet it at the points η, θ, ι, respectively, and join ηθ, θι, ηι. It is evident that ηθ, θι are sides of the hexagon inscribed in the larger circle. Now, on ηι let there be described a segment similar to that cut off by ηθ. Since, then, the square on ηι is necessarily three times greater than that on ηθ, the side of the hexagon,[50] and the square on ηθ six times that on αβ, it is evident that the segment described over ηι must be equal to the sum of the segments of the outer circle over ηθ and θι, together with those cut off in the inner circle by all the sides of the hexagon. If we now add, on both sides, the part of the triangle ηθι lying over

[48] A pentagon with a re-entrant angle is considered here: but observe—1°, that it is not called a pentagon, that term being then restricted to the regular pentagon; and, 2°, that it is described as a rectilineal figure composed of three triangles.

[49] It is assumed here that the angle βκε is obtuse, which it evidently is.

Bretschneider points out that in this paragraph the Greek text in the Aldine is corrupt, and consequently obscure: he corrects it by means of some transpositions and a few trifling additions. (See *Geom. vor Eukl.*, p. 118, note 2.)

[50] Then follows the proof of this statement, which I have omitted, as I think

the segment ηι, we arrive at the result that the triangle ηθι is equal to the lune ηθι, together with the segments of the inner circle cut off by the sides of the hexagon; and if we add on both sides the hexagon itself, we have the triangle,

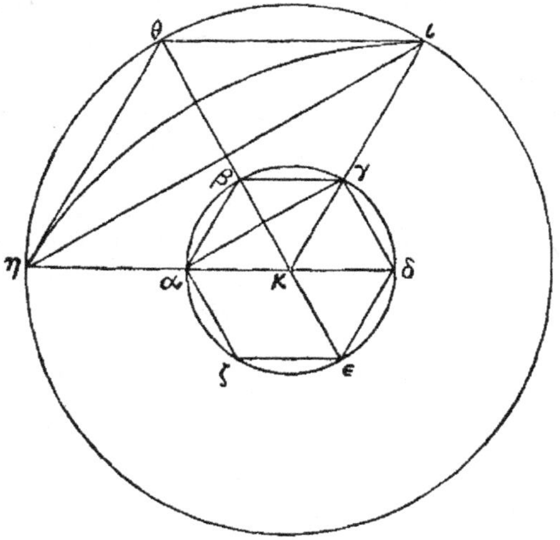

together with the hexagon, equal to the said lune together with the interior circle. Since, then, these rectilineal figures can be squared, the circle, together with the lune, can also be squared."

'Simplicius adds, in conclusion, that it must be admitted that Eudemus knows better all about Hippocrates of Chios, being nearer to him in point of time, and being also a pupil of Aristotle.'

If we examine this oldest fragment of Greek geometry, we see, in the first place, that there is in it a definition of similar segments of circles; they are defined to be those which contain the same quotum of their respective circles, as for instance, a semicircle is similar to a semi-

it was added by Simplicius: the word ἡ ὑποτείνουσα could scarcely have been used by Eudemus in the sense of sub-tense, as it is in this passage. Plato (Timaeus, 54, D, ed. Stallbaum, vol. VII., p. 228) and Aristotle use it, as we do, for the hypotenuse. It was sometimes used by later writers, Pappus for example, more generally, as it is here.

circle, the third part of one circle is similar to the third part of another circle.

Next we find the following theorems :—

(*a*). Similar segments contain equal angles ; [51]

(*b*). These in all semicircles are right ; segments which are larger or smaller than semicircles contain, respectively, acute or obtuse angles ;

(*c*). The side of a hexagon inscribed in a circle is equal to the radius ;

(*d*). In any triangle the square on a side opposite to an acute angle is less than the sum of the squares on the sides which contain the acute angle ;

(*e*). In an obtuse-angled triangle the square on the side subtending the obtuse angle is greater than the sum of the squares on the sides containing it ;

(*f*). In an isosceles triangle whose vertical angle is double the angle of an equilateral triangle, the square on the base is equal to three times the square on one of the equal sides ;

(*g*). In equiangular triangles the sides about the equal angles are proportional ;

(*h*). Circles are to each other as the squares on their diameters ;

(*i*). Similar segments of circles are to each other as the squares on their bases.

Lastly, we observe that the solution of the following problems is required :—

(*a*). Construct a square which shall be equal to a given rectilineal figure ;

(*b*). Find a line the square on which shall be equal to three times the square on a given line ; [52]

(*c*). Find a line such that twice the square on it shall be equal to three times the square on a given line ;

[51] For this, or rather its converse, is assumed in the demonstration, p. 73. Also, see p. 69.

[52] See theorem (*f*), *supra*.

(*d*). Being given two straight lines, construct a trapezium such that one of the parallel sides shall be equal to the greater of the two given lines, and each of the three remaining sides equal to the less;

(*e*). About the trapezium so constructed describe a circle;

(*f*). Describe a circle about a given triangle;

(*g*). From the extremity of the diameter of a semicircle draw a chord such that the part of it intercepted between the circle and a straight line drawn at right angles to the diameter at the distance of one half the radius shall be equal to a given straight line;

(*h*). Describe on a given straight line a segment of a circle which shall be similar to a given one.

There remain to us but few more notices of the work done by the geometers of this period:—

Antiphon, whose attempt to square the circle is given by Simplicius in the above extract, and who is also mentioned by Aristotle and some of his other commentators, is most probably the Sophist of that name who, we are told, often disputed with Socrates.[53] It appears from a notice of Themistius, that Antiphon started not only from the square, but also from the equilateral triangle, inscribed in a circle, and pursued the method and train of reasoning above described.[54]

Aristotle and his commentators mention another Sophist who attempted to square the circle—Bryson, of whom we have no certain knowledge, but who was probably a Pythagorean, and may have been the Bryson who is mentioned by Iamblichus amongst the disciples of Pythagoras.[55] Bryson inscribed a square,[56] or more generally

[53] Xenophon, *Memorab.* I., vi., § 1; Diog. Laert. II., 46, ed. Cobet, p. 44.

[54] Themist., f. 16; Brandis, *Schol. in Arist.*, p. 327^b, 33.

[55] Iambl., *Vit. Pyth.*, Cap. XXIII., 104.

[56] Alex. Aphrod., f. 30; Brandis, *Schol.*, p. 306^b.

any polygon,[57] in a circle, and circumscribed another of the same number of sides about the circle; he then argued that the circle is larger than the inscribed and less than the circumscribed polygon, and erroneously assumed that the excess in one case is equal to the defect in the other; he concluded thence that the circle is the mean between the two.

It seems, too, that some persons who had no knowledge of geometry took up the question, and fancied, as Alexander Aphrodisius tells us, that they should find the square of the circle in surface measure if they could find a square number which is also a cyclical number[58]—numbers as 5 or 6, whose square ends with the same number, are called by arithmeticians cyclical numbers.[59] On this Hankel observes that 'unfortunately we cannot assume that this solution of the squaring of the circle was only a joke'; and he adds, in a note, that 'perhaps it was of later origin, although it strongly reminds us of the Sophists who proved also that Homer's poetry was a geometrical figure because it is a circle of myths.'[60]

That the problem was one of public interest at that time, and that, further, owing to the false solutions of pretended geometers, an element of ridicule had become attached to it, is plain from the reference which Aristophanes makes to it in one of his comedies.[61]

In the last chapter, p. 28 (*t*), we saw that there was a tradition that the problem of the quadrature of the circle engaged the attention of the Pythagoreans. We saw, too (p. 47), that they probably derived the problem from the Egyptians, who sought to find from the diameter the side of a square whose area should be equal

57 Themist., f. 5; Brandis, *Schol.*, p. 211; Johan. Philop., f. 18; Brandis, *Schol.*, pp. 211, 212.

58 Simplicius, in Bretsch., *Geom. vor Eukl.*, p. 106.

59 *Ibid.*

60 Hankel, *Geschich. der Math.*, p. 116, and note.

61 *Birds*, 1005.

to that of the circle. From their approximate solution, it follows that the Egyptians must have assumed as evident that the area of a circle is proportional to the square on its diameter, though they would not have expressed themselves in this abstract manner. Anaxagoras (499–428 B.C.) is recorded to have investigated this problem during his imprisonment.[62]

Vitruvius tells us that Agatharchus invented scene-painting, and that he painted a scene for a tragedy which Æschylus brought out at Athens, and that he left notes on the subject. Vitruvius goes on to say that Democritus and Anaxagoras, profiting by these instructions, wrote on perspective.[63]

We have named DEMOCRITUS more than once: it is remarkable that the name of this great philosopher, who was no less eminent as a mathematician,[64] and whose fame stood so high in antiquity, does not occur in the summary of the history of geometry preserved by Proclus. In connection with this, we should note that Aristoxenus, in his *Historic Commentaries*, says that Plato wished to burn all the writings of Democritus that he was able to collect; but that the Pythagoreans, Amyclas and Cleinias, prevented him, as they said it would do no good, inasmuch as copies of his books were already in many hands. Diogenes Laertius goes on to say that it is plain that this was the case; for Plato, who mentions nearly all the ancient philosophers, nowhere speaks of Democritus.[65]

We are also told by Diogenes Laertius that Democritus was a pupil of Leucippus and of Anaxagoras, who was forty years his senior;[66] and further, that he went to

[62] Ἀλλ' Ἀναξαγόρας μὲν ἐν τῷ δεσμωτηρίῳ τὸν τοῦ κύκλου τετραγωνισμὸν ἔγραφε.—Plut., *de Exil.*, c. XVII., vol. III., p. 734, ed. Didot.

[63] *De Arch.*, VII., Praef.

[64] Cicero, *de finibus bonorum et malorum*, I., c. vi.; Diog. Laert., IX., vii., ed. Cobet, p. 236.

[65] Diog. Laert., *ibid.*, ed. Cobet, p. 237.

[66] *Ibid.*, p. 235.

Egypt to see the priests there, and to learn geometry from them.[67]

This report is confirmed by what Democritus himself tells us: 'I have wandered over a larger portion of the earth than any man of my time, inquiring about things most remote; I have observed very many climates and lands, and have listened to very many learned men; but no one has ever yet surpassed me in the construction of lines with demonstration; no, not even the Egyptian Harpedonaptae, as they are called (*καὶ γραμμέων συνθέσιος μετὰ ἀποδέξιος οὐδείς κώ με παρήλλαξε, οὐδ' οἱ Αἰγυπτίων καλεόμενοι Ἀρπεδονάπται*), with whom I lived five years in all, in a foreign land.'[68]

We learn further, from Diogenes Laertius, that Democritus was an admirer of the Pythagoreans; that he seems to have derived all his doctrines from Pythagoras, to such a degree, that one would have thought that he had been his pupil, if the difference of time did not prevent it; that at all events he was a pupil of some of the Pythagorean schools, and that he was intimate with (*συγγεγονέναι*) Philolaus.[69]

Diogenes Laertius gives a list of his writings: amongst those on mathematics we observe the following:—

Περὶ διαφορῆς γνώμονος ἢ περὶ ψαύσιος κύκλου καὶ σφαίρης (lit., On the difference of the gnomon, or on the contact of the circle and the sphere. Can what he has in view be the following idea: that, the gnomon, or carpenter's rule, being placed with its vertex on the circumference of a circle, in the limiting position, when one leg passes through the centre, the other will determine the tangent?); one on geometry; one on numbers; one on incommen-

[67] Diog. Laert., IX., vii., ed. Cobet, p. 236.

[68] Democrit., ap. Clem. Alex., *Strom.*, I., p. 304, ed. Sylburg; Mullach, *Fragm. Phil. Graec.*, p. 370.

[69] Diog. Laert., IX., vii., ed. Cobet, p. 236.

surable lines and solids, in two books; Ἀκτινογραφίη (a description of rays, probably perspective).[70]

We also learn, from a notice of Plutarch, that Democritus raised the following question: 'If a cone were cut by a plane parallel to its base [obviously meaning, what we should now call one infinitely near to that plane], what must we think of the surfaces of the sections, that they are equal or unequal? For if they are unequal, they will show the cone to be irregular, as having many indentations like steps, and unevennesses; and if they are equal, the sections will be equal, and the cone will appear to have the property of a cylinder, viz., to be composed of equal, and not unequal, circles, which is very absurd.'[71]

If we examine the contents of the foregoing extracts, and compare the state of geometry as presented to us in them with its condition about half a century earlier, we observe that the chief progress made in the interval concerns the circle. The early Pythagoreans seem not to have given much consideration to the properties of the circle; but the attention of the geometers of this period was naturally directed to them in connection with the problem of its quadrature.

We have already set down, *seriatim*, the theorems and problems relating to the circle which are contained in the extract from Eudemus.

Although the attempts of Antiphon and Bryson to square the circle did not meet with much favour from the ancient geometers, and were condemned on account of the paralogisms in them, yet their conceptions contain the first germ of the infinitesimal method: to Antiphon is due the merit of having first got into the right track by introducing for the solution of this problem—in accordance with the atomic theory then nascent—the fundamental

70 Diog. Laert., IX., vii., ed. Cobet, pp. 238, 239.

71 Plut., *de Comm. Not.*, vol. IV., p. 1321, ed. Didot.

idea of infinitesimals, and by trying to exhaust the circle by means of inscribed polygons of continually increasing number of sides; Bryson is entitled to praise for having seen the necessity of taking into consideration the circumscribed as well as the inscribed polygon, and thereby obtaining a superior as well as an inferior limit to the area of the circle. Bryson's idea is just, and should be regarded as complementary to the idea of Antiphon, which it limits and renders precise. Later, after the method of exhaustions had been invented, in order to supply demonstrations which were perfectly rigorous, the two limits, inferior and superior, were always considered together, as we see in Euclid and Archimedes.

We see, too, that the question which Plutarch tells us that Democritus himself raised involves the idea of infinitesimals; and it is evident that this question, taken in connection with the axiom in p. 57, must have presented real difficulties to the ancient geometers. The general question which underlies it was, as is well known, considered and answered by Leibnitz: 'Caeterùm aequalia esse puto, non tantùm quorum differentia est omnino nulla, sed et quorum differentia est incomparabiliter parva; et licèt ea Nihil omnino dici non debeat, non tamen est quantitas comparabilis cum ipsis, quorum est differentia. Quemadmodum si lineae punctum alterius lineae addas, vel superficiei lineam, quantitatem non auges. Idem est, si lineam quidem lineae addas, sed incomparabiliter minorem. Nec ulla constructione tale augmentum exhiberi potest. Scilicet eas tantùm homogeneas quantitates comparabiles esse, cum *Euclide, lib.* v., *defin.* 5, censeo, quarum una numero, sed finito, multiplicata, alteram superare potest. Et quae tali quantitate non differunt, aequalia esse statuo, quod etiam *Archimedes* sumsit, aliique post ipsum omnes. Et hoc ipsum est, quod dicitur differentiam esse data quavis minorem. Et *Archimedeo* quidem processu res semper deductione ad absurdum confirmari potest.'[72] Further, we have seen that Democritus wrote on the contact of the circle and of the sphere. The employment of the *gnomon* for the solution of this problem seems to show that Democritus, in its treatment, made use of the infinitesimal method; he might have employed the gnomon either in the manner indicated above, or, by making one leg of the gnomon pass through the centre of the circle, and moving the other parallel to itself, he could have found the middle points of a system of parallel chords, and thus ultimately the tangents parallel to them. At any rate this problem was a natural subject of inquiry for the chief founder of the atomic theory, just as Leibnitz—the author of the doctrine of monads and the founder of the infinitesimal calculus—was occupied with this same subject of tangency.

We observe, further, that the conception of the irrational (ἄλογον), which had been a secret of the Pythagorean school, became generally known, and that Democritus wrote a treatise on the subject.

We have seen that Anaxagoras and Democritus wrote on perspective, and that this is not the only instance in which the consideration of problems in geometry of three dimensions occupied the attention of Democritus.

On the whole, then, we find that considerable progress had been made in elementary geometry; and indeed the appearance of a treatise on the elements is in itself an indication of the same thing. We have further evidence of this, too, in the endeavours of the geometers of this period to extend to the circle and to volumes the results which had been arrived at concerning rectilineal figures and their comparison with each other. The Pythagoreans, as we have seen, had shown how to determine a square whose area was any multiple of a given square. The question now was to extend this to the cube, and, in

[72] Leibnitii, *Opera Omnia*, ed. L. Dutens, tom. III., p. 328.

particular, to solve the problem of the duplication of the cube.

Proclus (after Eudemus) and Eratosthenes tell us (*h* and *i*, p. 59) that Hippocrates reduced this question to one of plane geometry, namely, the finding of two mean proportionals between two given straight lines, the greater of which is double the less. Hippocrates, therefore, must have known that if four straight lines are in continued proportion, the first has the same ratio to the fourth that the cube described on the first, as side, has to the cube described in like manner on the second. He must then have pursued the following train of reasoning:—Suppose the problem solved, and that a cube is found which is double the given cube; find a third proportional to the sides of the two cubes, and then find a fourth proportional to these three lines; the fourth proportional must be double the side of the given cube; if, then, two mean proportionals can be found between the side of the given cube and a line whose length is double of that side, the problem will be solved. As the Pythagoreans had already solved the problem of finding a mean proportional between two given lines—or, which comes to the same, to construct a square which shall be equal to a given rectangle—it was not unreasonable for Hippocrates to suppose that he had put the problem of the duplication of the cube in a fair way of solution. Thus arose the famous problem of finding two mean proportionals between two given lines—a problem which occupied the attention of geometers for many centuries. Although, as Eratosthenes observed, the difficulty is not in this way got over; and although the new problem cannot be solved by means of the straight line and circle, or, in the language of the ancients, cannot be referred to plane problems, yet Hippocrates is entitled to much credit for this reduction of a problem in stereometry to one in plane geometry. The tragedy to which Eratosthenes refers in this account of the legendary origin of the problem is, according to Valckenaer, a lost play of Euripides, named Πολύειδος :[73] if this be so, it follows that this problem of the duplication of the cube, as well as that of the quadrature of the circle, was famous at Athens at this period.

Eratosthenes, in his letter to Ptolemy III., relates that one of the old tragic poets introduced Minos on the stage erecting a tomb for his son Glaucus; and then, deeming the structure too mean for a royal tomb, he said 'double it, but preserve the cubical form': *μικρόν γ' ἔλεξας βασιλεικοῦ σηκὸν τάφου, διπλάσιος ἔστω. τοῦ δὲ τοῦ κύβου μὴ σφαλείς.*[74] Eratosthenes then relates the part taken by Hippocrates of Chios towards the solution of this problem as given above (p. 59), and continues: 'Later [in the time of Plato], so the story goes, the Delians, who were suffering from a pestilence, being ordered by the oracle to double one of their altars, were thus placed in the same difficulty. They sent therefore to the geometers of the Academy, entreating them to solve the question.' This problem of the duplication of the cube—henceforth known as the *Delian Problem*—may have been originally suggested by the practical needs of architecture, as indicated in the legend, and have arisen in Theocratic times; it may subsequently have engaged the attention of the Pythagoreans as an object of theoretic interest and scientific inquiry, as suggested above.

These two ways of looking at the question seem suited for presenting it to the public on the one hand and to mathematical pupils on the other. From the consideration of a passage in Plutarch,[75] however, I am led to believe that the new problem—to find two mean proportionals

[73] See Reimer, *Historia problematis de cubi duplicatione*, p. 20, Gottingae, 1798; and Biering, *Historia problematis cubi duplicandi*, p. 6, Hauniae, 1844.

[74] Archim., ed. Torelli, p. 144. Valckenaer shows that these words of Eratosthenes contain two verses, which he thus restores:—

μικρὸν γ' ἔλεξας βασιλικοῦ σηκὸν τάφου·
διπλάσιος ἔστω, τοῦ κύβου δὲ μὴ σφαλῇς.

See Reimer, *l. c.*

[75] *Symp.*, VIII., *Quaestio* 2, c. iv.; Plut. *Opera*, ed. Didot, vol. IV., p. 277.

between two given lines—which arose out of it, had a deeper significance, and that it must have been regarded by the Pythagorean philosophers of this time as one of great importance, on account of its relation to their cosmology.

In the last chapter (p. 38) we saw that the Pythagoreans believed that the tetrahedron, octahedron, icosahedron, and cube corresponded to the four elements of the real world. This doctrine is ascribed by Plutarch to Pythagoras himself:[76] Philolaus, who lived at this time, also held that the elementary nature of bodies depended on their form. The tetrahedron was assigned to fire, the octahedron to air, the icosahedron to water, and the cube to earth; that is to say, it was held that the smallest constituent parts of these substances had each the form assigned to it.[77] This being so, what took place, according to this theory, when, under the action of heat, snow and ice melted, or water became vapour? In the former case, the elements which had been cubical took the icosahedral form, and in the latter the icosahedral elements became octahedral. Hence would naturally arise the following geometrical problems:—

Construct an icosahedron which shall be equal to a given cube;

Construct an octahedron which shall be equal to a given icosahedron.

Now Plutarch, in his *Symp.*, VIII., *Quaestio* ii.—Πῶς Πλάτων ἔλεγε τὸν θεὸν ἀεὶ γεωμετρεῖν, 3 & 4[78]—accepts this theory of

[76] Πυθαγόρας, πέντε σχημάτων ὄντων στερεῶν, ἅπερ καλεῖται καὶ μαθηματικά, ἐκ μὲν τοῦ κύβου φησὶ γεγονέναι τὴν γῆν, ἐκ δὲ τῆς πυραμίδος τὸ πῦρ, ἐκ δὲ τοῦ ὀκταέδρου τὸν ἀέρα, ἐκ δὲ τοῦ εἰκοσαέδρου τὸ ὕδωρ, ἐκ δὲ τοῦ δωδεκαέδρου τὴν τοῦ παντὸς σφαῖραν.

Πλάτων δὲ καὶ ἐν τούτοις πυθαγορίζει. Plut., *Plac.*, II., vi., 5 & 6; *Opera*, ed. Didot, vol. IV., p. 1081.

[77] Stob. *Eclog.* ab Heeren, lib. I., p. 10. See also Zeller, *die Philos. der Griechen*, Erster Theil, p. 376, Leipzig, 1876, *History of Greek Philosophy*, vol. I., p. 437, E. T.

[78] Plut. *Opera*, ed. Didot, vol. IV., pp. 876, 7.

Pythagoras and Philolaus, and in connection with it points out the importance of the problem: 'Given two figures, to construct a third which shall be equal to one of the two and similar to the other'—which he characterises as essentially geometrical, and attributes to Pythagoras (see Chapter II., p. 25). It is evident that Plutarch had in view solid and not plane figures; for, having previously referred to the forms of the constituent elements of bodies, viz. air, earth, fire, and water, as being those of the regular solids, omitting the dodecahedron, he goes on as follows: 'What,' said Diogenianus, 'has this [the problem—given two figures, to describe a third equal to one and similar to the other] to do with the subject?' 'You will easily know,' I said, 'if you call to mind the division in the Timaeus, which divided into three the things first existing, from which the Universe had its birth; the first of which three we call *God* [θεός, the arranger], a name most justly deserved; the second we call *matter*, and the third *ideal form*. . . . God was minded, then, to leave nothing, so far as it could be accomplished, undefined by limits, if it was capable of being defined by limits; but [rather] to adorn nature with proportion, measurement, and number: making some one thing [that is, the universe] out of the material taken all together; something that would be like the *ideal form*, and as big as the *matter*. So having given himself this problem, when the *two* were there, he made, and makes, and for ever maintains, a *third*, viz. the universe, which is equal to the *matter* and like the *model*.'

Let us now consider one of these problems—the former—and, applying to it the method of reduction, see what is required for its solution. Suppose the problem solved, and that an icosahedron has been constructed which shall be equal to a given cube. Take now another icosahedron, whose edge and volume are supposed to be known, and, pursuing the same method which was followed above in p. 84, we shall find that, in order to solve the problem, it would be necessary—

1. To find the volume of a polyhedron;

2. To find a line which shall have the same ratio to a given line that the volumes of two given polyhedra have to each other;

3. To find two mean proportionals between two given lines; and

4. To construct on a given line as edge a polyhedron which shall be similar to a given one.

Now we shall see that the problem of finding two mean proportionals between two given lines was first solved by Archytas of Tarentum—*ultimus Pythagoreorum*—then by his pupil Eudoxus of Cnidus, and thirdly by Menaechmus, who was a pupil of Eudoxus, and who used for its solution the conic sections which he had discovered: we shall see further that Eudoxus founded stereometry by showing that a triangular pyramid is one-third of a prism on the same base and between the same parallel planes; lastly, we shall find that these great discoveries were made with the aid of the method of geometrical analysis which either had meanwhile grown out of the method of reduction or was invented by Archytas.

It is probable that a third celebrated problem—the trisection of an angle—also occupied the attention of the geometers of this period. No doubt the Egyptians knew how to divide an angle, or an arc of a circle, into two equal parts; they may therefore have also known how to divide a right angle into three equal parts. We have seen, moreover, that the construction of the regular pentagon was known to Pythagoras, and we infer that he could have divided a right angle into five equal parts. In this way, then, the problem of the trisection of any angle—or the more general one of dividing an angle into any number of equal parts—would naturally arise. Further, if we examine the two reductions of the problem of the trisection of an angle which have been handed down to us from ancient times, we shall see that they are such as might naturally occur to the early geometers, and that they were quite within the reach of a Pythagorean—one who had worthily gone through his noviciate of at least two years of mathematical study and silent meditation. For this reason, and because, moreover, they furnish good examples of the method called ἀπαγωγή, I give them here.

Let us examine what is required for the trisection of an angle according to the method handed down to us by Pappus.[79]

Since we can trisect a right angle, it follows that the trisection of any angle can be effected if we can trisect an acute angle.

Let now $\alpha\beta\gamma$ be the given acute angle which it is required to trisect.

From any point α on the line $\alpha\beta$, which forms one leg of the given angle, let fall a perpendicular $\alpha\gamma$ on the other

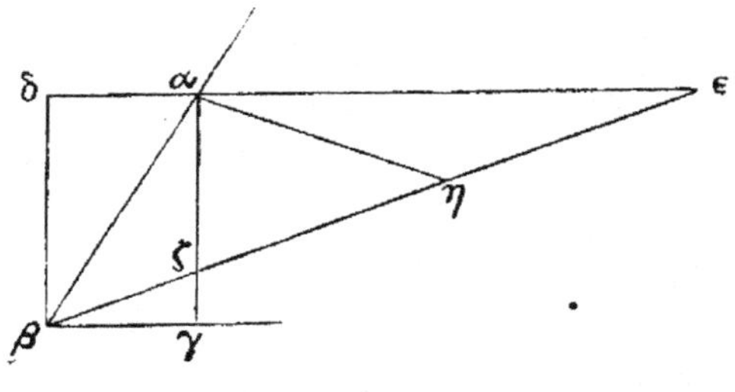

leg, and complete the rectangle $\alpha\gamma\beta\delta$. Suppose now that the problem is solved, and that a line is drawn making with $\beta\gamma$ an angle which is the third part of the given angle $\alpha\beta\gamma$; let this line cut $\alpha\gamma$ in ζ, and be produced until it meet $\delta\alpha$ produced at the point ϵ. Let now the straight line $\zeta\epsilon$ be bisected in η, and $\alpha\eta$ be joined; then the lines $\zeta\eta$, $\eta\epsilon$, $\alpha\eta$, and $\beta\alpha$ are all evidently equal to each other, and, therefore, the line $\zeta\epsilon$ is double of the line $\alpha\beta$, which is known.

The problem of the trisection of an angle is thus reduced to another:—

79 Pappi Alex., *Collect.*, ed. Hultsch, vol. I., p. 274.

From any vertex β of a rectangle $\beta\delta\alpha\gamma$ draw a line $\beta\zeta\epsilon$, so that the part $\zeta\epsilon$ of it intercepted between the two opposite sides, one of which is produced, shall be equal to a given line.

This reduction of the problem must, I think, be referred to an early period: for Pappus[80] tells us that when the ancient geometers wished to cut a given rectilineal angle into three equal parts they were at a loss, inasmuch as the problem which they endeavoured to solve as a plane problem could not be solved thus, but belonged to the class called solid;[81] and, as they were not yet acquainted with the conic sections, they could not see their way: but, later, they trisected an angle by means of the conic sections. He then states the problem concerning a rectangle, to which the trisection of an angle has been just now reduced, and solves it by means of a hyperbola.

The conic sections, we know, were discovered by Menaechmus, a pupil of Eudoxus (409-356 B.C.), and the discovery may, therefore, be referred to the middle of the fourth century.

Another method of trisecting an angle is preserved in the works of Archimedes, being indicated in Prop. 8 of the Lemmata[82]—a book which is a translation into Latin from the Arabic. The Lemmata are referred to Archimedes by some writers, but they certainly could not have come from him in their present form, as his name

[80] Pappi Alex., *Collect.*, ed. Hultsch, vol. I., p, 270 *et seq.*

[81] The ancients distinguished three kinds of problems—*plane*, *solid*, and *linear*. Those which could be solved by means of straight lines and circles were called plane; and were justly so called, as the lines by which the problems of this kind could be solved have their origin *in plano*. Those problems whose solution is obtained by means of one or more conic sections were called solid, inasmuch as for their construction we must use the superficies of solid figures—to wit, the sections of a cone. A third kind, called linear, remains, which required for their solution curves of a higher order, such as spirals, quadratrices, conchoids, and cissoids. See Pappi, *Collect.*, ed. Hultsch, vol. I., pp. 54 and 270.

[82] Archim. ex recens. Torelli, p. 358.

is quoted in two of the Propositions. They may have been contained in a note-book compiled from various sources by some later Greek mathematician,[83] and this Proposition may have been handed down from ancient times.

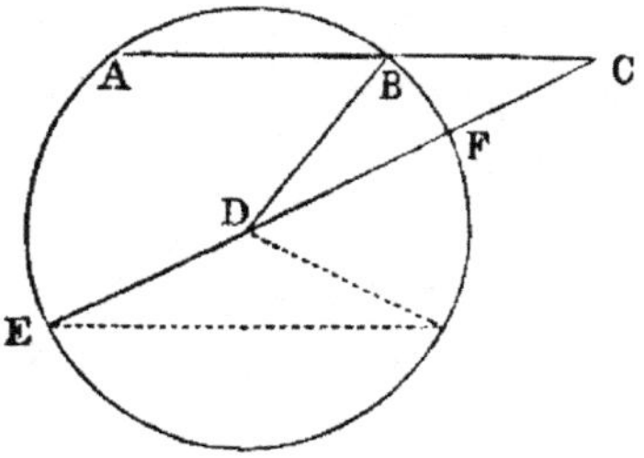

Prop. 8 of the Lemmata is: 'If a chord AB of a circle be produced until the part produced, BC, is equal to the radius; if then the point C be joined to the centre of the circle, which is the point D, and if CD, which cuts the circle in F, be produced until it cut it again in E, the arc AE will be three times the arc BF.' This theorem suggests the following reduction of the problem :—

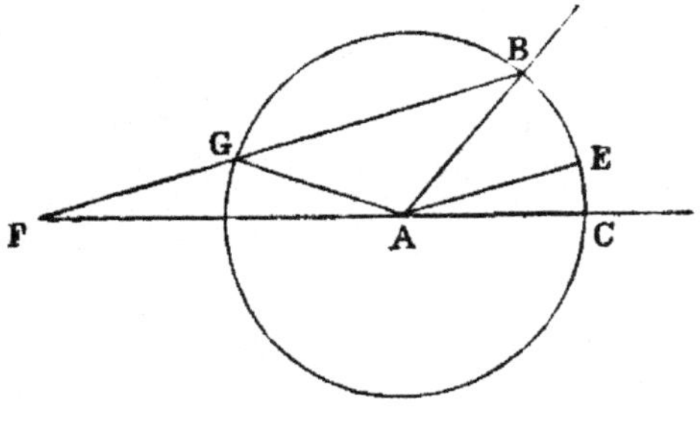

With the vertex A of the given angle BAC as centre, and any lines AC or AB as radius, let a circle be described. Suppose now that the problem is solved, and

[83] See Archim. ed. Torelli, *Praefatio*, pp. xviii and xix. See also Heiberg, *Quaestiones Archimedeae*, p. 24, who says: 'Itaque puto haec lemmata e plurium mathematicorum operibus esse excerpta, neque definiri jam potest, quantum ex iis Archimedi tribuendum sit.'

that the angle EAC is the third part of the angle BAC; through B let a straight line be drawn parallel to AE, and let it cut the circle again in G and the radius CA produced in F. Then, on account of the parallel lines AE and FGB, the angle ABG or the angle BGA, which is equal to it, will be double of the angle GFA; but the angle BGA is equal to the sum of the angles GFA and GAF; the angles GFA and GAF are, therefore, equal to each other, and consequently the lines GF and GA are also equal. The problem is, therefore, reduced to the following: From B draw the straight line BGF, so that the part of it, GF, intercepted between the circle and the diameter CAD produced shall be equal to the radius.[84]

For the reasons stated above, then, I think that the problem of the trisection of an angle was one of those which occupied the attention of the geometers of this period. Montucla, however, and after him many writers on the history of mathematics, attribute to Hippias of Elis, a contemporary of Socrates, the invention of a transcendental curve, known later as the Quadratrix of Deinostratus, by means of which an angle may be divided into any number of equal parts. This statement is made on the authority of the two following passages of Proclus:—

'Nicomedes trisected every rectilineal angle by means of the conchoidal lines, the inventor of whose particular nature he is, and the origin, construction, and properties of which he has explained. Others have solved the same problem by means of the quadratrices of Hippias and Nicomedes, making use of the mixed lines which are called quadratrices; others, again, starting from the spirals

[84] See F. Vietae, *Opera Mathematica*, studio F. à Schooten, p. 245, Lugd. Bat. 1646. These two reductions of the trisection of an angle were given by Montucla, but he did not give any references. See *Hist. des Math.*, tom. I., p. 194, 1iere ed.

of Archimedes, divided a rectilineal angle in a given ratio.[85].

'In the same manner other mathematicians are accustomed to treat of curved lines, explaining the properties of each form. Thus, Apollonius shows the properties of each of the conic sections; Nicomedes those of the conchoids; Hippias those of the quadratrix, and Perseus those of the spirics' (σπειρικῶν).[86]

Now the question arises whether the Hippias referred to in these two passages is Hippias of Elis. Montucla believes that there is some ground for this statement, for he says: 'Je ne crois pas que l'antiquité nous fournisse aucun autre géométre de ce nom, que celui dont je parle.'[87] Chasles, too, gives only a qualified assent to the statement. Arneth, Bretschneider, and Suter, however, attribute the invention of the quadratrix to Hippias of Elis without any qualification.[88] Hankel on the other hand, says that surely the Sophist Hippias of Elis cannot be the one referred to, but does not give any reason for his dissent.[89] I agree with Hankel for the following reasons:—

1. Hippias of Elis is not one of those to whom the progress of geometry is attributed in the summary of the history of geometry preserved by Proclus, although he is mentioned in it as an authority for the statement concerning Ameristus [or Mamercus].[90] The omission of his name would be strange if he were the inventor of the quadratrix.

85 Proclus, ed. Friedlein, p. 272.

86 *Ibid.*, p. 356.

87 Montucla, *Hist. des Math.*, tom. I., p. 181, nouvle. ed.

88 Chasles, *Hist. de la Géom.*, p. 8; Arneth, *Gesch. der Math.*, p. 95; Bretsch., *Geom. vor Eukl.*, p. 94; Suter, *Gesch. der Math. Wissenschaft.*, p. 32.

89 Hankel, *Gesch. der Math.*, p. 151, note. Hankel, also, in a review of Suter, *Geschichte der Mathematischen Wissenschaften*, published in the *Bullettino di Bibliografia e di Storia delle Scienze Matematiche e Fisiche*, says: 'A pag. 31 (lin. 3-6), Hippias, l'inventore della quadratrice, è identificato col Sofista Hippias, il che veramente avea già fatto il Bretschneider (pag. 94, lin. 39-42), ma senza darne la minima prova.' *Bullet.*, &c., tom. V., p. 297.

90 Proclus, ed. Friedlein, p. 65.

2. Diogenes Laertius tells us that Archytas was the first to apply an organic motion to a geometrical diagram;[91] and the description of the quadratrix requires such a motion.

3. Pappus tells us that: 'For the quadrature of a circle a certain line was assumed by Deinostratus, Nicomedes, and some other more recent geometers, which received its name from this property; it is called by them the quadratrix.'[92]

4. With respect to the observation of Montucla, I may mention that there was a skilful mechanician and geometer named Hippias contemporary with Lucian, who describes a bath constructed by him.[93]

I agree, then, with Hankel, that the invention of the quadratrix is erroneously attributed to Hippias of Elis. But Hankel himself, on the other hand, is guilty of a still

[91] Diog. Laert., VIII., c. iv., ed. Cobet, p. 224.

[92] Pappi, *Collect.*, ed. Hultsch, vol. I., pp. 250 and 252.

[93] *Hippias, seu Balneum.* Since the above was written I find that Cantor, *Vorles. über Gesch. der Math.*, p. 165, *sq.*, agrees with Montucla in this. He says: 'It has indeed been sometimes doubted whether the Hippias referred to by Proclus is really Hippias of Elis, but certainly without good grounds.' In support of his view Cantor advances the following reasons:—

1. Proclus in his commentary follows a custom from which he never deviates—he introduces an author whom he quotes with distinct names and surnames, but afterwards omits the latter when it can be done without an injury to distinctness. Cantor gives instances of this practice, and adds: 'If, then, Proclus mentions a Hippias, it must be Hippias of Elis, who had been already once distinctly so named in his *Commentary*.'

2. Waiving, however, this custom of Proclus, it is plain that with any author, especially with one who had devoted such earnest study to the works of Plato, Hippias without any further name could be only Hippias of Elis.

3. Cantor, having quoted passages from the dialogues of Plato, says: 'We think we may assume that Hippias of Elis must have enjoyed reputation as a teacher of mathematics at least equal to that which he had as a Sophist proper, and that he possessed all the knowledge of his time in natural sciences, astronomy, and mathematics.'

4. Lastly, Cantor tries to reconcile the passage quoted from Pappus with the two passages from Proclus: 'Hippias of Elis discovered about 420 B.C. a curve which could serve a double purpose—trisecting an angle and squaring the circle. From the latter application it got its name, *Quadratrix* (the Latin translation), but this name does not seem to reach further back than Deinostratus.'

greater anachronism in referring back the Method of Exhaustions to Hippocrates of Chios. He does so on grounds which in my judgment are quite insufficient. Hankel, after quoting from Archimedes the axiom—'If two spaces are unequal, it is possible to add their difference to itself so often that every finite space can be surpassed,' *see* p. 57—quotes further: 'Also, former geometers have made use of this lemma; for the theorem that circles are in the ratio of the squares of their diameters, &c., has been proved by the help of it. But each of the theorems mentioned is by no means less entitled to be accepted than those which have been proved without the help of that lemma; and, therefore, that which I now publish must likewise be accepted.' Hankel then reasons thus: 'Since, then, Archimedes brings this lemma into such connection with the theorem concerning the ratio of the areas of circles, and, on the other hand, Eudemus states that this theorem had been discovered and proved by Hippocrates, we may also assume that Hippocrates laid down the above axiom, which was taken up again by Archimedes, and which, in one shape or another, forms the basis of the Method of Exhaustions of the Ancients, *i.e.* of the method to exhaust, by means of inscribed and circumscribed polygons, the surface of a curvilinear figure. For this method necessarily requires such a principle in order to show that the curvilinear figure is really exhausted by these polygons.'[94] Eudemus, no doubt, stated that Hippocrates showed that circles have the same ratio as the squares on their diameters, but he does not give any indication as to the way in which the theorem was proved. An examination, however, of the portion of the passage quoted from Archimedes which is omitted by Hankel will, I think, show that there is no ground for his assumption.

The passage, which occurs in the letter of Archimedes to Dositheus prefixed to his treatise on the quadrature of

[94] Hankel, *Gesch. der Math.*, pp. 121–2.

the parabola, runs thus: 'Former geometers have also used this axiom. For, by making use of it, they proved that circles have to each other the duplicate ratio of their diameters; and that spheres have to each other the triplicate ratio of their diameters; moreover, that any pyramid is the third part of a prism which has the same base and the same altitude as the pyramid; also, that any cone is the third part of a cylinder which has the same base and the same altitude as the cone; all these they proved by assuming the axiom which has been set forth.'[95]

We see now that Archimedes does not bring this axiom into close connection with the theorem concerning the ratios of the areas of circles alone, but with three other theorems also; and we know that Archimedes, in a subsequent letter to the same Dositheus, which accompanied his treatise on the sphere and cylinder, states the two latter theorems, and says expressly that they were discovered by Eudoxus.[96] We know, too, that the doctrine of proportion, as contained in the Fifth Book of Euclid, is attributed to Eudoxus.[97] Further, we shall find that the invention of rigorous proofs for theorems such as Euclid, VI. 1, involves, in the case of incommensurable quantities, the same difficulty which is met with in proving rigorously the four theorems stated by Archimedes in connection with this axiom; and that in fact they all required a new method of reasoning—the Method of Exhaustions—which must, therefore, be attributed to Eudoxus.

The discovery of Hippocrates, which forms the basis of his investigation concerning the quadrature of the circle, has attracted much attention, and it may be interesting to

[95] Archim. ex recens. Torelli, p. 18.

[96] *Ibid.*, p. 64.

[97] We are told so in the anonymous scholium on the Elements of Euclid, which Knoche attributes to Proclus: see Eucl., *Elem.*, Graece ed. ab. August, pars. II, p. 329; also *Untersuchungen*, &c., Von Dr. J. H. Knoche, p. 10. Cf. p. 49, and note 76, *supra*.

inquire how it might probably have been arrived at. It appears to me that it might have been suggested in the following way:—Hippocrates might have met with the annexed figure, excluding the dotted lines, in the arts of decoration; and, contemplating the figure, he might have completed the four smaller circles and drawn their diameters, thus forming a square inscribed in the larger circle, as in the diagram. A diameter of the larger circle being then a diagonal of the square, whose sides are the diameters of the smaller circles, it follows that the larger circle is equal to the sum of two of the smaller circles. The larger circle is, therefore, equal to the sum of the four semicircles included by the dotted lines. Taking away the common parts—sc. the four segments of the larger circle standing on the sides of the square—we see that the square is equal to the sum of the four lunes.

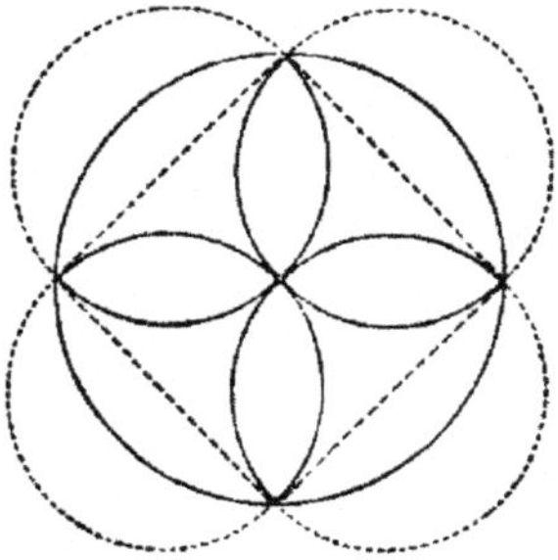

This observation—concerning, as it does, the geometry of areas—might even have been made by the Egyptians, who knew the geometrical facts on which it is founded, and who were celebrated for their skill in geometrical constructions. See Ch. II., pp. 29, 47, note 72.

In the investigation of Hippocrates given above we meet with manifest traces of an analytical method, as stated in Ch. II., p. 41, note 62. Indeed, Aristotle—and this is remarkable—after having defined ἀπαγωγή, evidently refers to a part of this investigation as an instance of it: for he says: 'Or again [there is reduction], if the middle terms between γ and β are few; for thus also there is a nearer approach to knowledge. For example, if δ were quadrature, and ϵ a rectilineal figure, and ζ a circle; if there were only one middle term between ϵ and ζ, viz., that a circle with lunes is equal to a rectilineal figure, there would be an approach to knowledge.'[98] See pp. 67, 68, above.

In many instances I have had occasion to refer to the method of reduction as one by which the ancient geometers made their discoveries, but perhaps I should notice that in general it was used along with geometrical constructions:[99] the importance attached to these may be seen from the passages quoted above from Proclus and Democritus, pp. 59, 80; as also from the fact that the Greeks had a special name, **ψευδογράφημα**, for a faulty construction.

The principal figure, then, amongst the geometers of this period is Hippocrates of Chios, who seems to have attracted notice as well by the strangeness of his career as by his striking discovery of the quadrature of the lune. Though his contributions to geometry, which have been set forth at length above, are in many respects important, yet the judgment pronounced on him by the ancients is certainly, on the whole, not a favourable one—witness the statements of Aristotle, Eudemus, Iamblichus, and Eutocius.

How is this to be explained? The faulty reasoning into which he is reported to have fallen in his pretended quadrature of the circle does not by itself seem to me to be a sufficient explanation of it: and indeed it is difficult

[98] ἢ πάλιν [ἀπαγωγή ἐστι] εἰ ὀλίγα τὰ μέσα τῶν ΒΓ· καὶ γὰρ οὕτως ἐγγύτερον τοῦ εἰδέναι. οἷον εἰ τὸ Δ εἴη τετραγωνίζεσθαι, τὸ δ' ἐφ' ᾧ Ε εὐθύγραμμον, τὸ δ' ἐφ' ᾧ Ζ κύκλος· εἰ τοῦ ΕΖ ἓν μόνον εἴη μέσον, τὸ μετὰ μηνίσκων ἴσον γίνεσθαι εὐθυγράμμῳ τὸν κύκλον, ἐγγὺς ἂν εἴη τοῦ εἰδέναι. *Anal. Prior.* II. xxv., p. 69[a], ed. Bekker. Observe the expressions τὸ δ' ἐφ' ᾧ Ε εὐθύγραμμον, &c., here, and see p. 72, note 45.

[99] Concerning the importance of *geometrical constructions* as a process of deduction, see P. Laffitte, *les Grands Types de l'Humanité*, vol. II., p. 329.

to reconcile such a gross mistake with the sagacity shown in his other discoveries, as Montucla has remarked.[100]

The account of the matter seems to me to be simply this:—Hippocrates, after having been engaged in commerce, went to Athens and frequented the schools of the philosophers—evidently Pythagorean—as related above. Now we must bear in mind that the early Pythagoreans did not commit any of their doctrines to writing[101]—their teaching being oral: and we must remember, further, that their pupils (ἀκουστικοί) were taught mathematics for several years, during which time a constant and intense application to the investigation of difficult questions was enjoined on them, as also silence—the rule being so stringent that they were not even permitted to ask questions concerning the difficulties which they met with:[102] and that after they had satisfied these conditions they passed into the class of mathematicians (μαθηματικοί), being freed from the obligation of silence; and it is probable that they then taught in their turn.

Taking all these circumstances into consideration, we may, I think, fairly assume that Hippocrates imperfectly understood some of the matter to which he had listened; and that, later, when he published what he had learned, he did not faithfully render what had been communicated to him.

If we adopt this view, we shall have the explanation of—

1. The intimate connection that exists between the work of Hippocrates and that of the Pythagoreans;

2. The paralogism into which he fell in his attempt to square the circle: for the quadrature of the lune on the side of the inscribed square may have been exhibited in the school, and then it may have been shown that the

100 Montucla, *Histoire des recherches sur la Quadrature du Cercle*, p. 39, nouvle. ed., Paris, 1831.

101 See Ch. II., p. 21, note 11, and the references given there.

102 See A. Ed. Chaignet, *Pythagore et la Philosophie Pythagoricienne*, vol. I., p. 115, Paris, 1874; see also Iambl., *Vit. Pyth.*, Cap. XVI., 68.

problem of the quadrature of the circle was reducible to that of the lune on the side of the inscribed hexagon; and what was stated conditionally may have been taken up by Hippocrates as unconditional;[103]

3. The further attempt which Hippocrates made to solve the problem by squaring a lune and circle together (see p. 74);

4. The obscurity and deficiency in the construction given in p. 72; and the dependence of that construction on a problem which we know was Pythagorean (see p. 24 (*e*), and note 26);[104]

5. The passage in Iamblichus, see p. 58 (*f*); and, generally, the unfavourable opinion entertained by the ancients of Hippocrates.

This conjecture gains additional strength from the fact that the publication of the Pythagorean doctrines was first made by Philolaus, who was a contemporary of Socrates, and, therefore, somewhat junior to Hippocrates: Philolaus may have thought that it was full time to make this publication, notwithstanding the Pythagorean precept to the contrary.

The view which I have taken of the form of the

[103] In reference to this paralogism of Hippocrates, Bretschneider (*Geom. vor Eukl.*, p. 122) says, 'It is difficult to assume so gross a mistake on the part of such a good geometer,' and he ascribes the supposed error to a complete misunderstanding. He then gives an explanation similar to that given above, with this difference, that he supposes Hippocrates to have stated the matter correctly, and that Aristotle took it up erroneously: it seems to me more probable that Hippocrates took up wrongly what he had heard at lecture than that Aristotle did so on reading the work of Hippocrates. Further, we see from the quotation in p. 98, from *Anal. Prior.*, that Aristotle fully understood the conditions of the question.

[104] Referring to *the application of areas*, Mr. Charles Taylor, *An Introduction to the Ancient and Modern Geometry of Conics*, Prolegomena, p. xxv., says, 'Although it has not been made out wherein consisted the importance of the discovery in the hands of the Pythagoreans, we shall see that it played a great part in the system of Apollonius, and that he was led to designate the three conic sections by the Pythagorean terms Parabola, Hyperbola, Ellipse.'

I may notice that we have an instance of these problems in the construction referred to above: for other applications of the method see Ch. II., pp. 41, 43.

demonstrations in geometry at this period differs altogether from that put forward by Bretschneider and Hankel, and agrees better not only with what Simplicius tells us 'of the summary manner of Eudemus, who, according to archaic custom, gives concise proofs' (see p. 69), but also with what we know of the origin, development, and transmission of geometry: as to the last, what room would there be for the silent meditation on difficult questions which was enjoined on the pupils in the Pythagorean schools, if the steps were minute, and if laboured proofs were given of the simplest theorems?

The need of a change in the method of proof was brought about at this very time, and was in great measure due to the action of the Sophists, who questioned everything.

Flaws, no doubt, were found in many demonstrations which had hitherto passed current; new conceptions arose, while others, which had been secret, became generally known, and gave rise to unexpected difficulties; new problems, whose solution could not be effected by the old methods, came to the front, and attracted general attention. It became necessary then on the one hand to recast the old methods, and on the other to invent new methods, which would enable geometers to solve the new problems.

I have already indicated the men who were equal to this task, and I propose in the following chapters to examine their work.

CHAPTER IV.*

ARCHYTAS.

State of Hellas during the last generation of the fifth century B.C.—Magna Graecia again became flourishing.—Archytas of Tarentum.—His life, eminence as a Statesman and noble character.—Notices of his Geometrical work.—Was there a Roman Agrimensor named Architas?—The problem 'to find two Mean Proportionals between two Given Lines' was first solved by Archytas.—His Solution.—Theorems which occur in it.—Inferences from it as to Archytas's knowledge of Geometry.—The conception of Geometrical Loci involved in this Solution.—Different opinions as to its importance.—Construction of Archytas's Solution.—Was Plato the inventor of the method of Geometrical Analysis?—Passage in the 'Republic' of Plato, in which the backward state of Solid Geometry is noticed.—Yet Archytas had, for the period, a profound knowledge of Geometry of Three Dimensions; and Stereometry was founded in Plato's lifetime by Eudoxus.

DURING the last thirty years of the fifth century before the Christian era no progress was made in geometry at Athens, owing to the Peloponnesian war, which having broken out between the two principal States of Greece, gradually spread to the other States, and for the space of a generation involved almost the whole of Hellas. Although it was at Syracuse that the issue was really decided, yet the Hellenic cities of Italy kept aloof from the contest,[1] and Magna Graecia enjoyed at this time a

* In the preparation of this and the following Chapters I have again made use of the works of Bretschneider and Hankel, and have derived much advantage from the great work of Cantor—*Vorlesungen über Geschichte der Mathematik.* I have also constantly used the *Index Graecitatis* appended by Hultsch to vol. III. of his edition of Pappus; which, indeed, I have found invaluable.

[1] At the time of the Athenian expedition to Sicily they were not received into any of the Italian cities, nor were they allowed any market, but had only the liberty of anchorage and water—and even that was denied them at Tarentum and Locri. At Rhegium, however, though the Athenians were not received into the city, they were allowed a market without the walls; they then made proposals to the Rhegians, begging them, as Chalcideans, to aid the Leontines. 'To which was answered, that they would take part with neither, but whatever should seem fitting to the rest of the Italians that they also would do.' Thucyd. VI. 44.

period of comparative rest, and again became flourishing. This proved to be an event of the highest importance; for, some years before the commencement of the Peloponnesian war, the disorder which had long prevailed in the cities of Magna Graecia had been allayed through the intervention of the Achaeans;[2] party feeling, which had run so high, had been soothed, and the banished Pythagoreans allowed to return. The foundation of Thurii (443 B.C.), under the auspices of Pericles, in which the different Hellenic races joined, and which seems not to have incurred any opposition from the native tribes, may be regarded as an indication of the improved state of affairs, and as a pledge for the future.[3] It is probable that the pacification was effected by the Achaeans on condition

[2] 'The political creed and peculiar form of government now mentioned also existed among the Achaeans in former times. This is clear from many other facts, but one or two selected proofs will suffice, for the present, to make the thing believed. At the time when the Senate-houses (συνέδρια) of the Pythagoreans were burnt in the parts about Italy then called Magna Graecia, and a universal change of the form of government was subsequently made (as was likely when all the most eminent men in each State had been so unexpectedly cut off), it came to pass that the Grecian cities in those parts were inundated with bloodshed, sedition, and every kind of disorder. And when embassies came from very many parts of Greece with a view to effect a cessation of differences in the various States, the latter agreed in employing the Achaeans, and their well-known integrity, for the removal of existing evils. Not only at this time did they adopt the system of the Achaeans, but, some time after, they set about imitating their form of government in a complete and thorough manner. For the people of Crotona, Sybaris, and Caulon, sent for them by common consent; and first of all they established a common temple dedicated to Zeus, 'the Giver of Concord,' and a place in which they held their meetings and deliberations: in the second place, they took the customs and laws of the Achaeans, and applied themselves to their use, and to the management of their public affairs in accordance with them. But some time after, being hindered by the overbearing power of Dionysius of Syracuse, and also by the encroachments made upon them by the neigbouring natives of the country, they renounced them, not voluntarily, but of necessity.' Polybius, II., 39. Polybius uses συνέδριον for the senate at Rome: there would be one in each Graeco-Italian State—a point which, as will be seen, has not been sufficiently noted.

[3] The foundation of Thurii, near the site of Sybaris, seems to have been regarded as an event of high importance; Herodotus was amongst the first citizens, and Empedocles visited Thurii soon after it was founded. The names of the tribes of Thurii show the pan-Hellenic character of the foundation.

that, on the one hand, the banished Pythagoreans should be allowed to return to their homes, and, on the other, that they should give up all organised political action.[4] Whether this be so or not, many Pythagoreans returned to Italy, and the Brotherhood ceased for ever to exist as a political association.[5] Pythagoreanism, thus purified, continued as a religious society and as a philosophic School; further, owing to this purification and to the members being thus enabled to give their undivided atten-

[4] Chaignet, *Pythagore et la Philosophie Pythagoricienne*, I., p. 93, says so, but does not give his authority; the passage in Polybius, II. 39, to which he refers, does not contain this statement.

[5] There are so many conflicting accounts of the events referred to here that it is impossible to reconcile them (see p. 53). The view which I have adopted seems to me to fit best with the contemporary history, with the history of geometry, and with the balance of the authorities. Zeller, on the other hand, thinks that the most probable account is 'that the first public outbreak must have taken place after the death of Pythagoras, though an opposition to him and his friends may perhaps have arisen during his lifetime, and caused his migration to Metapontum. The party struggles with the Pythagoreans, thus begun, may have repeated themselves at different times in the cities of Magna Graecia, and the variations in the statements may be partially accounted for as recollections of these different facts. The burning of the assembled Pythagoreans in Crotona, and the general assault upon the Pythagorean party, most likely did not take place until the middle of the fifth century; and lastly, Pythagoras may have spent the last portion of his life unmolested at Metapontum.' (Zeller, *Pre-Socratic Philosophy*, vol. I., p. 360, E. T.).

Ueberweg takes a similar view:—

'But the persecutions were also several times renewed. In Crotona, as it appears, the partisans of Pythagoras and the 'Cylonians' were for a long time after the death of Pythagoras living in opposition as political parties, till at length, about a century later, the Pythagoreans were surprised by their opponents, while engaged in a deliberation in the 'house of Milo' (who himself had died long before), and the house being set on fire and surrounded, all perished with the exception of Archippus and Lysis of Tarentum. (According to other accounts, the burning of the house, in which the Pythagoreans were assembled, took place on the occasion of the first reaction against the Society, in the lifetime of Pythagoras.) Lysis went to Thebes, and was there (soon after 400 B.C.) a teacher of the youthful Epaminondas.' (Ueberweg, *History of Philosophy*, vol. I., p. 46, E. T.)

Zeller, in a note on the passage quoted above, gives the reasons on which his suppositions are chiefly based. Chaignet, *Pyth. et la Phil. Pyth.* vol. I., p. 88, and note, states Zeller's opinion, and, while admitting that the reasons advanced by him do not want force, says that they are not strong enough to convince him: he then gives his objections. Chaignet, further on, p. 94, *n.*, referring to the name Italian, by which the Pythagorean philosophy is known, says: 'C'est même ce qui tion and their whole energy to the solution of scientific questions, it became as distinguised and flourishing as ever: at this time, too, remarkable instances of devoted friendship and of elevation of character are recorded of some of the body. Towards the end of this and the beginning of the following centuries encroachments were made on the more southerly cities by the native populations, and some of them were attacked and taken by the elder Dionysius:[6] meanwhile Tarentum, provided with an excellent

me fait croire que les luttes intestines n'ont pas eu la durée que suppose M. Zeller; car si les pythagoriciens avaient été exilés pendant près de soixante dix ans de l'Italie, comment le nom de l'Italie serait-il devenu ou resté attaché à leur école?' Referring to this objection of Chaignet, Zeller says, 'I know not with what eyes he can have read a discussion which expressly attempts to show that the Pythagoreans were not expelled till 440, and returned before 406' (*loc cit.* p. 363, note).

To the objections urged by Chaignet I would add—

1. Nearly all agree in attributing the origin of the troubles in Lower Italy to the events which followed the destruction of Sybaris.

2. The fortunes of Magna Graecia seem to have been at their lowest ebb at the time of the Persian war; this appears from the fact that, before the battle of Salamis, ambassadors were sent by the Lacedemonians and Athenians to Syracuse and Corcyra, to invite them to join the defensive league against the Persians, but passed by Lower Italy.

3. The revival of trade consequent on the formation of the confederacy of Delos, 476 B.C., for the protection of the Aegean Sea, must have had a beneficial influence on the cities of Magna Graecia, and the foundation of Thurii, 443 B.C., is in itself an indication that the settlement of the country had been already effected.

4. The answer of the Rhegians to Nicias, 415 B.C., shows that at that time there existed a good understanding between the Italiot cities.

5. Zeller's argument chiefly rests on the assumption that Lysis, the teacher of Epaminondas, was the same as the Lysis who in nearly all the statements is mentioned along with Archippus as being the only Pythagoreans who escaped the slaughter. Bentley had long ago suggested that they were not the same. Lysis and Archippus are mentioned as having handed down Pythagorean lore as heirlooms in their families (Porphyry, *Vit. Pyth.*, p. 101, Didot). This fact is in my judgment decisive of the matter; for when Lysis, the teacher of Epaminondas, lived there were no longer any secrets. See p. 22, *n.* 11.

[6] In 393 B.C. a league was formed by some of the cities in order to protect themselves against the Lucanians and against Dionysius. Tarentum appears not to have joined the league till later, and then its colony Heraclea was the place of meeting. The passage in Thucydides, quoted above, shows, however, that long before that date a good understanding existed between the cities of Magna Graecia.

harbour, and, on account of its remote situation, not yet threatened, had gained in importance, and was now the most opulent and powerful city in Magna Graecia. In this city, at this time, Archytas—the last great Pythagorean—grew to manhod.

ARCHYTAS of Tarentum[7] was a contemporary of Plato (428–347 B.C.), but probably senior to him, and was said by some to have been one of Plato's Pythagorean teachers[8] when he visited Italy. Their friendship[9] was proverbial, and it was he who saved Plato's life when he was in danger of being put to death by the younger Dionysius (about 361 B.C.). Archytas was probably, almost certainly, a pupil of Philolaus.[10] We have the following particulars of his life:—

He was a great statesman, and was seven times[11] appointed general of his fellow-citizens, notwithstanding the law which forbade the command to be held for more than one year, and he was, moreover, chosen commander-in-chief, with autocratic powers, by the confederation of the Hellenic cities of Magna Graecia;[12] it is further stated that

[7] See Diog. Laert. VIII. c. iv. See also J. Navarro, *Tentamen de Archytae Tarentini vita atque operibus*, Pars Prior. Hafniae, 1819, and authorities given by him.

[8] Cic *de Fin.* V., xxix. 87; *Rep.* I. 10, 16; *de Senec.* 12, 41. Val. Max. VIII. 7.

[9] Iambl., *Vit. Pyth.* 127, p. 48, ed. Didot. 'Verum ergo illud est quod, a Tarentino Archyta, ut opinor, dici solitum, nostros senes commemorare audivi ab aliis senibus auditum: "si quis in caelum ascendisset naturamque mundi et pulchritudinem siderum perspexisset, insuavem illam admirationem ei fore, quae jucundissuma fuisset, si aliquem cui narraret habuisset." Sic natura solitarium nihil amat, semperque ad aliquod tamquam adminiculum adnititur quod in amicissimo quoque dulcissimum est.'—Cic., *de Amic.* 23, 87.

[10] Cic., *de Oratore*, Lib. III. xxxiv. 139, aut *Philolaus Archytam Tarentinum?* The common reading *Philolaum Archytas Tarentinus*, which is manifestly wrong, was corrected by Orellius.

[11] Diog. Laert. *loc. cit.* Ælian, *Var. Hist.* VII. 14, says *six*.

[12] Τοῦ κοινοῦ δὲ τῶν Ἰταλιωτῶν προέστη, στρατηγὸς αἱρεθεὶς αὐτοκράτωρ ὑπὸ τῶν πολιτῶν καὶ τῶν περὶ ἐκεῖνον τὸν τόπον Ἑλλήνων. Suidas, *sub v.* This title στρατηγὸς αὐτοκράτωρ was conferred on Nicias and his colleagues by the Athenians when they sent their great expedition to Sicily: it was also conferred

he was never defeated as a general, but that, having once given up his command through being envied, the troops he had commanded were at once taken prisoners: he was celebrated for his domestic virtues, and several touching anecdotes are preserved of his just dealings with his slaves, and of his kindness to them and to children.[13] Aristotle even mentions with praise a toy that was invented by him for the amusement of infants:[14] he was the object of universal admiration on account of his being endowed with every virtue;[15] and Horace, in a beautiful Ode,[16] in which he refers to the death of Archytas by shipwreck in the Adriatic Sea, recognises his eminence as an arithmetician, geometer, and astronomer.

In the list of works written by Aristotle, but unfortunately lost, we find three books on the philosophy of Archytas, and one [*τὰ ἐκ τοῦ Τιμαίου καὶ τῶν Ἀρχυτείων ά*]; these, however, may have been part of his works[17] on the Pythagoreans which occur in the same list, but which also are lost. Some works attributed to Archytas have come down to us, but their authenticity has been questioned, especially by Grüppe, and is still a matter of dispute:[18] these works, however, do not concern geometry.

He is mentioned by Eudemus in the passage quoted from Proclus in the Introduction (p. 4.) along with his contemporaries, Leodamas of Thasos and Theaetetus of

by the Syracusans on the elder Dionysius: Diodorus, XIII. 94. See Arnold, *History of Rome*, I. p. 448, *n.* 18.

[13] As to the former, which was in accordance with Pythagorean principles, see Iambl., *Vit. Pyth.* XXXI. 197, pp. 66, 67, ed. Did.; Plutarch, *de ed. puer.* III., p. 12, ed. Did.; as to the latter, see Athenaeus, XII. 16; Aelian, *Var. hist.* XII. 15.

[14] Aristot. Pol. V. (8), c. vi. See also Suidas.

[15] ἐθαυμάζετο δὲ καὶ παρὰ τοῖς πολλοῖς ἐπὶ πάσῃ ἀρετῇ, Diog. Laert. *loc. cit.*

[16] I. 28.

[17] Diog. Laert. V. i., ed. Cobet, p. 116. This, however, could hardly have been so, as *one* book only on the Pythagoreans is mentioned, and *one* against them.

[18] Grüppe, *ueber die Fragmente des Archytas und der älteren Pythagoreer.* Berlin, 1840.

Athens, who were also contemporaries of Plato, as having increased the number of demonstrations of theorems and solutions of problems, and developed them into a larger and more systematic body of knowledge.[19]

· The services of Archytas, in relation to the doctrine of proportion, which are mentioned in conjunction with those of Hippasus and Eudoxus, have been noticed in pp. 27 (*o*) and 45.

One of the two methods of finding right-angled triangles whose sides can be expressed by numbers—the Platonic one, namely, which sets out from even numbers—is ascribed to Architas [no doubt, Archytas of Tarentum] by Boethius:[20] see pp. 34, 35, and note 53. I have there given the two rules of Pythagoras and Plato for finding right-angled triangles, whose sides can be expressed by numbers; and I have shown how the method of Pythagoras, which sets out from odd numbers, results at once from the consideration of the formation of squares by the addition of consecutive gnomons, each of which contains an odd number of squares. I have shown, further, that the method attributed to Plato by Heron and Proclus, which proceeds from even numbers, is a simple and natural extension of the method of Pythagoras: indeed it is difficult to conceive that an extension so simple and natural could have escaped the notice of his successors. Now Aristotle tells us that Plato followed the Pythagoreans in many things;[21] Alexander Aphrodisiensis, in

[19] Proclus, ed. Friedlein, p. 66.

[20] Boet., *Geom.*, ed. Friedlein, p. 408. Heiberg, in a notice of Cantor's 'History of Mathematics,' *Revue Critique d'Histoire et de Littérature*, 16 Mai, 1881, pp. 378, 9, remarks, 'Il est difficile de croire à l'existence d'un auteur romain nommé Architas, qui aurait écrit sur l'arithmétique, et dont le nom, qui ne serait, du reste, ni grec ni latin, aurait totalement disparu avec ses œuvres, à l'exception de quelques passages dans Boèce.' The question, however, still remains as to the authenticity of the *Ars Geometriae*. Cantor stoutly maintains that the *Geometry* of Boethius is genuine: Friedlein, the editor of the edition quoted, on the other hand dissents; and the great majority of philologists agree in regarding the question as still *sub judice*. See *Rev. Crit.* loc. cit.

[21] Arist., *Met.* I. 6, p. 987^a, ed. Bek.

his *Commentary* on the Metaphysics, repeats this statement;[22] Asclepius goes further and says, not in many things but in everything.[23] Even Theon of Smyrna, a Platonist, in his work 'Concerning those things which in Mathematics are useful for the reading of Plato,' says that Plato in many places follows the Pythagoreans.[24] All this being considered, it seems to me to amount almost to a certainty that Plato learned his method for finding right-angled triangles whose sides can be expressed numerically from the Pythagoreans; he probably then introduced it into Greece, and thereby got the credit of having invented his rule. It follows also, I think, that the Architas referred to by Boethius could be no other than the great Pythagorean philosopher of Tarentum.

The belief in the existence of a Roman agrimensor named Architas, and that he was the man to whom Boethius—or the pseudo-Boethius—refers, is founded on a remarkable passage of the *Ars Geometriae*,[25] which, I think, has been incorrectly interpreted, and also on another passage in which Euclid is mentioned as prior to Architas.[26] The former passage, which is as follows:—'Sed jam tempus est ad geometricalis mensae traditionem ab Archita, non sordido hujus disciplinae auctore, Latio accommodatam venire, si prius praemisero,' &c., is translated by Cantor thus: 'But it is time to pass over to the communication of the geometrical table, which was prepared for Latium by Architas, no mean author of this science, when I shall first have mentioned,' &c.:[27] this, in my opinion, is not the sense of the passage. I think that 'ab Archita' should be taken with *traditionem*, and not with *accommo-*

22 Alex. Aph., *Schol. in Arist.*, Brand., p. 548ª, 8.
23 Asclep., *Schol. loc. cit.*, p. 548ª, 35.
24 Theon. Smyrn. *Arithm.*, ed. de Gelder, p. 17.
25 Boet., ed. Friedlein, p. 393.
26 *Id.*, p. 412.
27 Cantor, *Gesch. der Math.*, p. 493.

datam, the correct translation being—'But it is now time to come to the account of the geometrical table as given by Architas ("no mean authority" in this branch of learning), as adapted by me to Latin readers; when,' &c. Now it is remarkable—and this, as far as I know, has been overlooked—that the author of the *Ars Geometriae*, whoever he may have been, applies to Architas the very expression applied by Archytas to Pythagoras in Hor. *Od.* I. 28:

> 'iudice te, non sordidus auctor
> 'naturae verique.'

The mention of Euclid as prior to Archytas is easily explained, since we know that for centuries Euclid the geometer was confounded with Euclid of Megara,[28] who was a contemporary of Archytas, but senior to him.

We learn from Diogenes Laertius that he was the first to employ scientific method in the treatment of Mechanics, by introducing the use of mathematical principles; and was also the first to apply a mechanical motion in the solution of a geometrical problem, while trying to find by means of the section of a semi-cylinder two mean proportionals, with a view to the duplication of the cube.[29]

Eratosthenes, too, in his letter to Ptolemy III., having

[28] This error seems to have originated with Valerius Maximus (VIII. 12), an author probably of the time of the emperor Tiberius, and was current in the middle ages.

[29] *οὗτος πρῶτος τὰ μηχανικὰ ταῖς μαθηματικαῖς προσχρησάμενος ἀρχαῖς μεθώδευσε, καὶ πρῶτος κίνησιν ὀργανικὴν διαγράμματι γεωμετρικῷ προσήγαγε, διὰ τῆς τομῆς τοῦ ἡμικυλίνδρου δύο μέσας ἀνὰ λόγον λαβεῖν ζητῶν εἰς τὸν τοῦ κύβου διπλασιασμόν.* Diog. Laert. *loc. cit.*, ed. Cobet., p. 224.

That is, he first propounded the affinity and connection of Mechanics and Mathematics with one another, by applying Mathematics to Mechanics, and mechanical motion to Mathematics.

This seems to be the meaning of the passage: but Mechanics, or rather Statics, was first raised to the rank of a demonstrative science by Archimedes, who founded it on the principle of the lever. Archytas, however, was a practical mechanician, and his wooden flying dove was the wonder of antiquity. Favorinus, see Aul. Gell., *Noctes Atticae*, X. 12.

related the origin of the Delian Problem (see p. 85), tells us that 'the Delians sent a deputation to the geometers who were staying with Plato at Academia, and requested them to solve the problem for them. While they were devoting themselves without stint of labour to the work, and trying to find two mean proportionals between the two given lines, Archytas of Tarentum is said to have discovered them by means of his semicylinders, and Eudoxus by means of the so called 'Curved Lines' (*διὰ τῶν καλουμένων καμπύλων γραμμῶν*). It was the lot, however, of all these men to be able to solve the problem with satisfactory demonstration; while it was impossible to apply their methods practically so that they should come into use; except, to some small extent and with difficulty, that of Menaechmus.'[30]

There is also a reference to this in the epigram which closes the letter of Eratosthenes.[31]

The solution of Archytas, to which these passages refer, has come down to us through Eutocius, and is as follows:—

'*The invention of Archytas as Eudemus relates it.*'[32]

'Let there be two given lines, $\alpha\delta$, γ; it is required to find two mean proportionals to them. Let a circle $\alpha\beta\delta\zeta$ be described round the greater line $\alpha\delta$; and let the line $\alpha\beta$, equal to γ, be inserted in it; and being produced let it meet at the point π, the line touching the circle at the

30 Archimedes, ex recens. Torelli, p. 144; Archimedis, *Opera Omnia*, ed. J. L. Heiberg, vol. III., pp. 104, 106.

31 μηδὲ σύ γ' Ἀρχύτεω δυσμήχανα ἔργα κυλίνδρων,
μηδὲ Μενεχμείους κωνοτομεῖν τριάδας
δίζηαι, μηδ' εἴ τι θεουδέος Εὐδόξοιο
καμπύλον ἐν γραμμαῖς εἶδος ἀναγράφεται.

Archim., ex. rec. Torelli, p. 146; Archim. *Opera*, ed. Heiberg, vol. III., p. 112.

32 *Ibid.*, ex. rec. Tor. p. 143; *Ibid.*, ed. Heib. vol. III., p. 98.

point δ: further let βεζ be drawn parallel to πδ. Now let it be conceived that a semicylinder is erected on the semicircle αβδ, at right angles to it: also, at right angles to it, let there be drawn on the line αδ a semicircle lying in the parallelogram of the cylinder. Then let this semicircle be turned round from the point δ towards β, the extremity α of the diameter remaining fixed; it will in its circuit cut the cylindrical surface and describe on it a certain line.

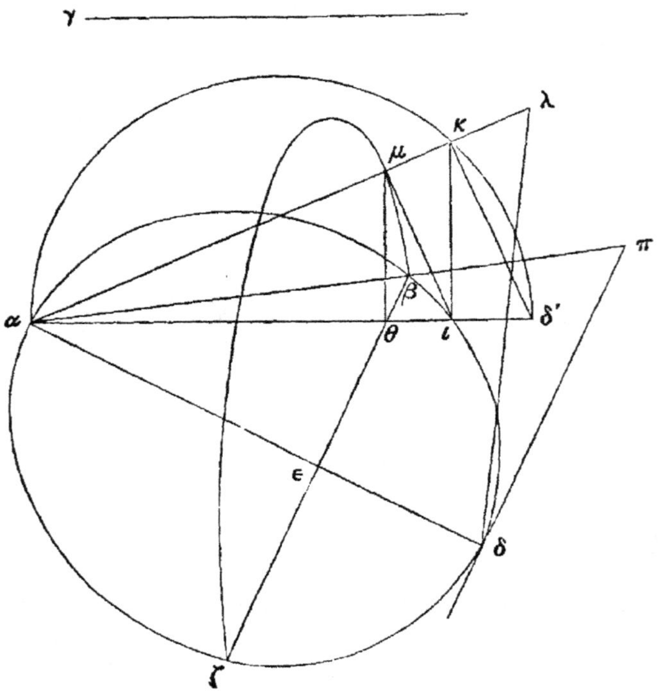

Again, if, the line αδ remaining fixed, the triangle απδ be turned round, with a motion contrary to that of the semicircle, it will form a conical surface with the straight line απ, which in its circuit will meet the cylindrical line [*i.e.* the line which is described on the cylindrical surface by the motion of the semicircle] in some point; at the same time the point β will describe a semicircle on the surface of the cone. Now, at the place[33] of meeting of the lines, let the semicircle in the course of its motion have a position $\delta'\kappa\alpha$, and the triangle in the course of its opposite motion a position $\delta\lambda\alpha$; and let the point of the said meeting be κ. Also let the semicircle described by β be $\beta\mu\zeta$, and the common section of it and of the circle $\beta\delta\zeta\alpha$ be $\beta\zeta$: now from the point κ let a perpendicular be drawn to the plane of the semicircle $\beta\delta\alpha$; it will fall on the periphery of the circle, because the cylinder stands perpendicularly. Let it fall, and let it be $\kappa\iota$; and let the line joining the points ι and α meet the line $\beta\zeta$ in the point θ; and let the right line $\alpha\lambda$ meet the semicircle $\beta\mu\zeta$ in the point μ; also let the lines $\kappa\delta'$, $\mu\iota$, $\mu\theta$ be drawn.

'Since, then, each of the semicircles $\delta'\kappa\alpha$, $\beta\mu\zeta$ is at right angles to the underlying plane, and, therefore, their common section $\mu\theta$ is at right angles to the plane of the circle; so also is the line $\mu\theta$ at right angles to $\beta\zeta$. Therefore, the rectangle under the lines $\theta\beta$, $\theta\zeta$; that is, under $\theta\alpha$, $\theta\iota$; is equal to the square on $\mu\theta$. The triangle $\alpha\mu\iota$ is therefore similar to each of the triangles $\mu\iota\theta$, $\mu\alpha\theta$, and the angle $\iota\mu\alpha$ is right. But the angle $\delta'\kappa\alpha$ is also right. Therefore, the lines $\kappa\delta'$, $\mu\iota$ are parallel. And there will be the proportion:—As the line $\delta'\alpha$ is to $\alpha\kappa$, *i.e.* $\kappa\alpha$ to $\alpha\iota$, so is the line $\iota\alpha$ to $\alpha\mu$, on account of the similarity of the triangles. The four straight lines $\delta'\alpha$, $\alpha\kappa$, $\alpha\iota$, $\alpha\mu$ are, therefore, in continued proportion. Also the line $\alpha\mu$ is equal to γ, since it is equal to the line $\alpha\beta$. So the two lines $\alpha\delta$, γ being given, two mean proportionals have been found, viz. $\alpha\kappa$, $\alpha\iota$.'

Although this extract from the *History of Geometry* of Eudemus seems to have been to some extent modernised by the omission of certain archaic expressions such as those referred to in the preceding chapter (p. 72, *n.* 45) [and by the introduction of the phrase 'parallelogram of

[33] ἐχέτω δὴ θέσιν κατὰ τὸν τόπον τῆς συμπτώσεως τῶν γραμμῶν τὸ μὲν κινούμενον ἡμικύκλιον ὡς τὴν τοῦ ΔΚΑ., &c.

the cylinder'],* yet the whole passage appears to me to bear the impress of Eudemus's clear and concise style: further, it agrees perfectly with the report of Diogenes Laertius, and also with the words in the letter of Eratosthenes to Ptolemy III., which have been given above. If now we examine its contents, and compare them with those of the more ancient fragment, we shall find a remarkable progress.

The following theorems occur in it:—

(*a*). If a perpendicular be drawn from the vertex of a right-angled triangle on the hypotenuse, each side is a mean proportional between the hypotenuse and its adjacent segment.[34]

(*b*). The perpendicular is the mean proportional between the segments of the hypotenuse;[35] and, conversely, if the perpendicular on the base of a triangle be a mean proportional between the segments of the base, the vertical angle is right.

(*c*). If two chords of a circle cut one another, the rectangle under the segments of one is equal to the rectangle under the segments of the other. This was most probably obtained by similar triangles, and, therefore, required the following theorem, the ascription of which to Hippocrates has been questioned.

(*d*). The angles in the same segment of a circle are equal to each other.

(*e*). Two planes which are perpendicular to a third plane intersect in a line which is perpendicular to that plane, and also to their lines of intersection with the third plane.

Archytas, as we see from his solution, was familiar

* [The term parallelogram was invented by Euclid: see Proclus, ed. Friedlein, pp. 392, 3. Cf. Heiberg, *Litterargeschichtliche Studien über Euklid*, p. 35.]

[34] The whole investigation is, in fact, based on this theorem.

[35] The solution of the Delian problem attributed to Plato, and by Menaechmus, are founded on this theorem.

with the generation of cylinders and cones, and had also clear ideas on the interpenetration of surfaces; he had, moreover, a correct conception of geometrical loci, and of their application to the determination of a point by means of their intersection. Further, since by the theorem of Thales the point μ must lie on a semicircle of which $\alpha\iota$ is the diameter, we shall see hereafter that in the solution of Archytas the same conceptions are made use of and the same course of reasoning is pursued, which, in the hands of his successor and contemporary Menaechmus, led to the discovery of the three conic sections. Such knowledge and inventive power surely outweigh in importance many special theorems.

Cantor, indeed, misconceiving the sense of the word τόπος, supposes that the expression '*geometrical locus*' occurs in this passage. He says: 'In the text handed down by Eutocius, even the word τόπος, *geometrical locus*, occurs. If we knew with certainty that here Eutocius reports literally according to Eudemus, and Eudemus literally according to Archytas, this expression would be very remarkable, because it corresponds with an important mathematical conception, the beginnings of which we are indeed compelled to attribute to Archytas, whilst we find it hard to believe in a development of it at that time which has proceeded so far as to give it a name. In our opinion, therefore, Eudemus, who was probably followed very closely by Eutocius, allowed himself, in his report on the doubling of the cube by Archytas, some changes in the style, and in this manner the word "*locus*," which in the meanwhile had obtained the dignity of a technical term, has been inserted. This supposition is supported by the fact that the whole statement of the procedure of Archytas sounds far less antique than, for instance, that of the attempts at quadrature of Hippocrates of Chios. Of course we only assume that Eudemus has, to a certain extent, treated the wording of Archytas freely.

The sense he must have rendered faithfully, and thus the conclusions we have drawn as to the stereometrical knowledge of Archytas remain untouched.'[36]

This reasoning of Cantor is based on a misconception of the meaning of the passage in which the word τόπος occurs; τόπος in it merely means *place*, as translated above. Though Cantor's argument, founded on the occurrence of the word τόπος, is not sound; yet, as I have said, the solution of Archytas involves the conception of *geometrical loci*, and the determination of a point by means of their intersection—not merely 'the beginnings of the conception,' as Cantor supposes; for surely such a notion could not first arise with a curve of double curvature. The first beginning of this notion has been referred to Thales in the first chapter[37] (p. 13).

Further, Archytas makes use of the theorem of Thales—the angle in a semicircle is right. He shows, moreover, that $\mu\theta$ is a mean proportional between $\alpha\theta$ and $\theta\iota$, and concludes that the angle $\iota\mu\alpha$ is right: it seems to me, therefore, to be a fair inference from this that he must have seen that the point μ may lie anywhere on the circumference of a circle of which $\alpha\iota$ is the diameter. Now Eutocius, in his *Commentaries* on the *Conics* of Apollonius,[38] tells us what the old geometers meant by *Plane Loci*, and gives some examples of them, the first of which is this very theorem. It is as follows:—

[36] Cantor, *Gesch. der Math.*, p. 197.

[37] Speaking of the solution of the 'Delian Problem' by Menaechmus, Favaro observes: 'Avvertiamo espressamente che Menecmo non fu egli stesso l'inventore di questa dottrina [dei luoghi geometrici]. Montucla (*Histoire des Mathématiques*, nouvelle édition, tome premier, à Paris, An. VII., p. 171); e Chasles (*Aperçu historique*, Bruxelles, 1837, p. 5) la attribuiscono alla scuola di Platone; G. Johnston Allman (*Greek Geometry from Thales to Euclid*, Dublin, 1877, p. 171) la fa risalire a Talete, appoggiando la sua argumentazione con valide ragioni.' Antonio Favaro, *Notizie Storico-Critiche sulla Contruzione delle Equazioni*. Modena, 1878, p. 21.

[38] Apollonius, *Conic.*, ed. Halleius, p. 10.

'A finite straight line being given, to find a point from which the perpendicular drawn to the given line shall be a mean proportional between the segments. Geometers call such a point a *locus*, since not one point only is the solution of the problem, but the whole place which the circumference of a circle described on the given line as diameter occupies: for if a semicircle be described on the given line, whatever point you may take on the circumference, and draw from it a perpendicular on the diameter, that point will solve the problem.'

Eutocius then gives a second example—'A straight line being given, to find a point without it from which the straight lines drawn to its extremities shall be equal to each other'—on which he makes observations of a similar character, and then adds: 'To the same effect Apollonius himself writes in his *Locus Resolutus*, with the subjoined [figure]:

"'Two points in a plane being given, and the ratio of two unequal lines being also given, a circle can be described in the plane, so that the straight lines inflected from the given points to the circumference of the circle shall have the same ratio as the given one."'

Then follows the solution, which is accompanied with a diagram. As this passage is remarkable in many respects, I give the original:—

Τὸ δὲ τρίτον τῶν κωνικῶν περιέχει, φησὶ, πολλὰ καὶ παράδοξα θεωρήματα χρήσιμα πρὸς τὰς συνθέσεις τῶν στερεῶν τόπων. Ἐπιπέδους τόπους ἔθος τοῖς παλαιοῖς γεωμέτραις λεγειν, ὅτε τῶν προβλημάτων οὐκ ἀφ' ἑνὸς σημείου μόνον, ἀλλ' ἀπὸ πλειόνων γίνεται τὸ ποίημα· οἷον ἐν ἐπιτάξει, τῆς εὐθείας δοθείσης πεπερασμένης εὑρεῖν τι σημεῖον ἀφ' οὗ ἡ ἀχθεῖσα κάθετος ἐπὶ τὴν δοθεῖσαν μέση ἀνάλογον γίνεται τῶν τμημάτων. Τόπον καλοῦσι τὸ τοιοῦτον, οὐ μὸνον γὰρ ἓν σημεῖόν ἐστι τὸ ποιοῦν τὸ πρόβλημα, ἀλλὰ τόπος ὅλος ὃν ἔχει ἡ περιφέρεια τοῦ περὶ διάμετρον τὴν δοθεῖσαν εὐθεῖαν κύκλου· ἐὰν γὰρ ἐπὶ τῆς δοθείσης εὐθείας ἡμικύκλιον γραφῇ, ὅπερ ἂν ἐπὶ τῆς περιφερείας λάβῃς σημεῖον, καὶ ἀπ' αὐτοῦ κάθετον ἀγάγῃς ἐπι τὴν διάμετρον, ποιήσει τὸ προβληθέν . . . . ὅμοιον καὶ γράφει αὐτὸς Ἀπολλώνιος ἐν τῷ ἀναλυομένῳ τόπῳ, ἐπὶ τοῦ ὑποκειμένου.[39]

Δύο δοθέντων σημείων ἐν ἐπιπέδῳ καὶ λόγου δοθέντος ἀνίσων εὐθειῶν δυνατόν ἐστιν ἐν τῷ ἐπιπέδῳ γράψαι κύκλον ὥστε τὰς ἀπὸ τῶν δοθέντων σημείων ἐπὶ τὴν περιφέρειαν τοῦ κύκλου κλωμένας εὐθείας λόγον ἔχειν τὸν αὐτὸν τῷ δοθέντι.

It is to be observed, in the first place, that a contrast is here made between Apollonius and the old geometers (*οἱ παλαιοὶ γεωμέτραι*), the same expression which, in p. 90, we found was used by Pappus in speaking of the geometers before the time of Menaechmus. Secondly, on examination it will be seen that *loci*, as *e. g.* those given above, partake of a certain ambiguity, since they can be enunciated either as theorems or as problems; and we shall see later that, about the middle of the fourth century B. C., there was a discussion between Speusippus and the philosophers of the Academy on the one side, and Menaechmus, the pupil and, no doubt, successor of Eudoxus, and the mathematicians of the school of Cyzicus, on the other, as to whether everything was a theorem or everything a problem: the mathematicians, as might be expected, took the latter view, and the philosophers, just as naturally, held the former. Now it was to propositions of this ambiguous character that the term *porism*, in the sense in which it is now always used, was applied—a signification which was quite consistent with the etymology of the word.[40] Lastly, the reader will not fail to observe that the first of the three *loci* given above

[39] Heiberg, in his *Studien über Euklid*, p. 70, reads τὸ ὑποκείμενον, and adds in a note that Halley has ὑποκειμένῳ, in place of τὸ ὑποκείμενον, a statement which is not correct. I have interpreted Halley's reading as referring to the subjoined diagram.

[40] πορίζεσθαι, to procure. The question is—in a *theorem*, to *prove* something; in a *problem* to *construct* something; in a *porism*, to *find* something. So the conclusion of the theorem is, ὅπερ ἔδει δεῖξαι, Q. E. D., of the problem, ὅπερ ἔδει ποιῆσαι, Q. E. F., and of the porism, ὅπερ ἔδει εὑρεῖν, Q. E. I. Amongst the ancients the word *porism* had also another signification, that of corollary. See Heiberg, *Stud. über Eukl.*, pp. 56–79, where the obscure subject of *porisms* is treated with remarkable clearness.

is strikingly suggestive of the method of Analytic Geometry. As to the term τόπος, it may be noticed that Aristaeus, who was later than Menaechmus, but prior to Euclid, wrote five books on *Solid Loci* (οἱ στερεοὶ τόποι).[41] In conclusion, I cannot agree with Cantor's view that the passage has the appearance of being modernised in expression [with the exception of the use of the Euclidean term 'parallelogram']: there is nothing in the text from which any alteration in phraseology can be inferred, as there can be in the two solutions of the 'Delian Problem' by Menaechmus, in which the words parabola and hyperbola occur.

The solution of Archytas seems to me not to have been duly appreciated. Montucla does not give the solution, but refers to it in a loose manner, and says that it was merely a *geometrical curiosity*, and of no practical importance.[42] Chasles, who, as we have seen (p. 13, *n.* 19), in the History of Geometry before Euclid, copies Montucla, also says that the solution was purely speculative; he even gives an inaccurate description of the construction—taking an *arête* of the cylinder as axis of the cone[43]—in which he is followed by some more recent writers.[44] Flauti, on the other hand, gives a clear and full account of the method of Archytas, and shows how his solution may be actually constructed. For this purpose it is necessary to give a construction for finding the intersection of the surface of the semicylinder with that of the *tore* generated by the revolution of the semicircle round the *side* of the cylinder through the point a as axis; and also for finding the intersection of the surface of the same

[41] Pappi, *Collect.*, ed. Hultsch, vol. II., p. 672.

[42] 'Mais ce n'étoit-là qu'une curiosité géométrique, uniquement propre à satisfaire l'esprit, et dont la pratique ne sçauroit tirer aucun secours.'—Montucla, *Histoire des Math.*, tom. I., p. 188.

[43] Chasles, *Aperçu hist.*, p. 6.

[44] *e.g.* Hoefer, *Histoire des Math.*, p. 133.

semicylinder with that of the cone described by the revolution of the triangle $\alpha\pi\delta$: the intersection of these curves gives the point κ, and then the point ι, by means of which the problem is solved. Now, in order to determine the point κ, it will be sufficient to find the projections of these two curves on the vertical plane on $\alpha\delta$, which contains the axes of the three surfaces of revolution concerned, and which Archytas calls the parallelogram of the cylinder. The projection on this plane of the curve of intersection of the *tore* and semicylinder can be easily found: the projection of the point κ, for example, is at once obtained by drawing from the point ι, which is the projection of the point κ on the horizontal plane $\alpha\beta\delta$, a perpendicular $\iota\xi$ on $\alpha\delta$, and then at the point ξ erecting in the vertical plane a perpendicular $\xi\eta$ equal to $\iota\kappa$, the ordinate of the semicircle $\alpha\kappa\delta'$, corresponding to the point ι; and in like manner for all other points. The projection on the same vertical plane of the curve of intersection of the cone and semicylinder can also be found: for example, the projection of the point κ, which is the intersection of $\alpha\kappa$ and $\iota\kappa$, the sides of the cone and cylinder, on the vertical plane, is the intersection of the projections of these lines on that plane; the latter projection is the line $\xi\eta$, and the former is obtained by drawing in the vertical plane, through the point ϵ, a line $\epsilon\nu$ perpendicular to $\alpha\delta$ and equal to $\theta\mu$, the ordinate of the semicircle $\beta\mu\zeta$, and then joining $\alpha\nu$, and producing it to meet $\xi\eta$; and so for all other points on the curve of intersection of the cone and cylinder.[45] So far Flauti.

Each of these projections can be constructed by points:—

To find the ordinate of the first of these curves corresponding to any point ξ, we have only to describe a square, whose area is the excess of the rectangle under the line $\alpha\delta$ and a mean proportional between the lines $\alpha\delta$ and $\alpha\xi$, over the square on the mean: the side of this square is

[45] Flauti, *Geometria di Sito*, terza edizione. Napoli, 1842, pp. 192–194.

the ordinate required.[46] In order to describe the projection of the intersection of the cone and cylinder, it will be sufficient to find the length, $\alpha\xi$, which corresponds to any ordinate, $\xi\eta$ ($=\iota\kappa$), supposed known, of this curve; and to effect this we have only to apply to the given line $\alpha\epsilon$ a rectangle, which shall be equal to the square on the line $\xi\eta$, and which shall be *excessive* by a rectangle similar to a given one, namely, one whose sides are the lines $\alpha\delta$ and $\alpha\epsilon$—*i.e.* the greater of the two given lines, between which the two mean proportionals are sought, and the third proportional to it and the less.[47]

[46] For, $\xi\eta^2 = \iota\kappa^2 = \alpha\iota \,.\, \iota\delta' = \alpha\iota \,.\, (\alpha\delta' - \alpha\iota)$; but $\alpha\delta' = \alpha\delta$;

therefore, $\xi\eta^2 = \alpha\delta \,.\, \alpha\iota - \alpha\iota^2$. Again, since $\alpha\delta : \alpha\iota :: \alpha\iota : \alpha\xi$,

we have also $$\xi\eta^2 = \alpha\delta \,.\, (\sqrt{\alpha\delta \,.\, \alpha\xi} - \alpha\xi).$$

[47] Thus, $\theta\mu^2 = \beta\epsilon^2 - \theta\epsilon^2$. Now $\theta\mu = \epsilon\nu$, and $\epsilon\nu : \xi\eta :: \alpha\epsilon : \alpha\xi$;

we have, therefore, $$\epsilon\nu^2 = \frac{\alpha\epsilon^2 \,.\, \xi\eta^2}{\alpha\xi^2} = \beta\epsilon^2 - \theta\epsilon^2.$$

Hence, $$\xi\eta^2 = \frac{\beta\epsilon^2}{\alpha\epsilon^2} \,.\, \alpha\xi^2 - \frac{\theta\epsilon^2}{\alpha\epsilon^2} \,.\, \alpha\xi^2 = \frac{\beta\epsilon^2}{\alpha\epsilon^2} \,.\, \alpha\xi^2 - \iota\xi^2,$$

since $$\theta\epsilon : \iota\xi :: \alpha\epsilon : \alpha\xi.$$

But $$\iota\xi^2 = \alpha\xi \,.\, (\alpha\delta - \alpha\xi);$$

hence we get $$\xi\eta^2 = \frac{\alpha\beta^2}{\alpha\epsilon^2} \,.\, \alpha\xi^2 - \alpha\delta \,.\, \alpha\xi;$$

and, finally, since $$\alpha\delta : \alpha\beta :: \alpha\beta : \alpha\epsilon,$$

we have $$\xi\eta^2 = \frac{\alpha\delta^2}{\alpha\beta^2} \,.\, \alpha\xi^2 - \alpha\delta \,.\, \alpha\xi.$$

The equations of these projections can, as M. Paul Tannery has shown (*Sur les Solutions du Problème de Délos par Archytas et par Eudoxe*, Mémoires de la Société des Sciences physiques et naturelles de Bordeaux, 2ᵉ série, tome II., p. 277), be easily obtained by analytic geometry. Taking, as axes of co-ordinates, the line $\alpha\delta$, the tangent to the circle $\alpha\beta\delta$ at the point α, and the side of the cylinder through the point α, the equations of the three surfaces are:—

the cylinder, $$x^2 + y^2 = ax;$$

the tore, $$x^2 + y^2 + z^2 = a\sqrt{x^2 + y^2};$$

the cone, $$x^2 + y^2 + z^2 = \frac{a^2}{b^2}x^2,$$

where a and b are the lines $\alpha\delta$ and $\alpha\beta$, between which the two mean proportionals are sought.

So much ingenuity and ability are shown in the treatment of this problem by Archytas, that the investigation of these projections, in itself so natural,[48] seems to have been quite within his reach, especially as we know that the subject of Perspective had been treated of already by Anaxagoras and Democritus (see pp. 79, 83). It may be observed, further, that the construction of the first projection is easily obtained; and as to the construction of the second projection, we see that it requires merely the solution of a problem attributed to the Pythagoreans by Eudemus, simpler cases of which we have already met with (see p. 24 (*e*), and pp. 41, 72, *n.* 46). On the other hand, it should be noticed—1° that we do not know when the description of a curve by points was first made; 2° that the second projection, which is a hyperbola, was obtained later by Menaechmus as a section of the cone; 3° and lastly, that the names of the conic sections—*parabola*, *hyperbola*, and *ellipse*—derived from the problems concerning the *application*, *excess*, and *defect* of areas, were first given to them by Apollonius.[49]

We easily obtain from these three equations:—

$$x = b\sqrt[3]{\frac{b}{a}}; \qquad \sqrt{x^2+y^2} = \sqrt[3]{ab^2}; \qquad \text{and} \quad \sqrt{x^2+y^2+z^2} = \sqrt[3]{a^2b}.$$

The last two give the first and second mean proportionals between b and a.

We also obtain easily the projections on the plane of zx of the curves of intersection of the cylinder and tore—

$$z^2 = a\sqrt{x}\,(\sqrt{a} - \sqrt{x}),$$

and of the cylinder and cone,

$$z^2 = \frac{a^2}{b^2}x^2 - ax.$$

These results agree with those obtained above geometrically.

[48] 'La recherche des projections sur les plans donnés des intersections deux à deux des surfaces auxiliaires est, à cet égard, si naturelle que, si l'on peut s'étonner d'une chose, c'est précisément qu' Archytas ait conservé à sa solution une forme purement theorique.' P. Tannery, *loc. cit.*, p. 279.

[49] See *supra*, p. 24, and *n.* 26: see also Apollonii *Conica*, ed. Halleius, p. 9, also pp. 31, 33, 35; and Pappi *Collect.*, ed. Hultsch, vol. II., p. 674; and Proclus, ed. Friedlein, p. 419.

Several authors give Archytas credit for a knowledge of the geometry of space, which was quite exceptional and remarkable at that time, and they notice the peculiarity of his making use of a curve of double curvature —the first, as far as we know, conceived by any geometer; but no one, I believe, has pointed out the importance of the conceptions and method of Archytas in relation to the invention of the conic sections, and the filiation of ideas seems to me to have been completely overlooked.

Bretschneider, not bearing in mind what Simplicius tells us of Eudemus's concise proofs, thinks that this solution, though faithfully transmitted, may have been somewhat abbreviated. He thinks, too, that it must belong to the later age of Archytas—a long time after the opening of the Academy—inasmuch as the discussion of sections of solids by planes, and of their intersections with each other, must have made some progress before a geometer could have hit upon such a solution as this; and also because such a solution was, no doubt, possible only when Analysis was substituted for Synthesis.[50]

Bretschneider even attempts to detect the particular analysis by which Archytas arrived at his solution, and as Cantor thinks, with tolerable success.[51] The latter reason goes on the assumption, current since Montucla, that Plato was the inventor of the method of geometrical analysis—an assumption which is based on the following passages in Diogenes Laertius and Proclus:—

He [Plato] first taught Leodamas of Thasos the analytic method of inquiry.[52]

Methods are also handed down, of which the best is that through analysis, which brings back what is required

[50] Bretsch., *Geom. vor Eukl.*, pp. 151, 152.

[51] Cantor, *Gesch. der Math.*, p. 198.

[52] καὶ πρῶτος τὸν κατὰ τὴν ἀνάλυσιν τῆς ζητήσεως τρόπον εἰσηγήσατο Λεωδάμαντι τῷ Θασίῳ. Diog. Laert., III. 24, ed. Cobet, p. 74.

to some admitted principle, and which Plato, as they say, transmitted to Leodamas, who is reported to have become thereby the discoverer of many geometrical theorems.[53]

Some authors, on the other hand, think, and as it seems to me with justice, that these passages prove nothing more than that Plato communicated to Leodamas of Thasos this method of analysis with which he had become acquainted, most probably, in Cyrene and Italy.[54] It is to be remembered that Plato—who in mathematics seems to have been painstaking rather than inventive—has not treated of this method in any of his numerous writings, nor is he reported to have made any discoveries by means of it, as Leodamas and Eudoxus are said to have done, and as we know Archytas and Menaechmus did. Indeed we have only to compare the solution attributed to Plato of the problem of finding two mean proportionals—which must be regarded as purely mechanical, inasmuch as the geometrical theorem on which it is based is met with in the solution of Archytas—with the highly rational solutions of the same problem by Archytas and Menaechmus, to see the wide interval between them and him in a mathematical point of view. Plato, moreover, was the pupil of Socrates, who held such mean views of geometry as to say that it might be cultivated only so far as that a person might be able to distribute and accept a piece of land by measure.[55] We know that Plato, after his master's death, went to Cyrene

[53] Μέθοδοι δὲ ὅμως παραδίδονται· καλλίστη μὲν ἡ διὰ τῆς ἀναλύσεως ἐπ' ἀρχὴν ὁμολογουμένην ἀνάγουσα τὸ ζητούμενον, ἣν καὶ ὁ Πλάτων, ὥς φασι, Λεωδάμαντι παρέδωκεν. ἀφ' ἧς καὶ ἐκεῖνος πολλῶν κατὰ γεωμετρίαν εὑρετὴς ἱστόρηται γενέσθαι.—Proclus, ed. Friedlein, p. 211.

[54] J. J. de Gelder quotes these passages of Diogenes Laertius and Proclus, and adds: 'Haec satis testantur doctissimum Montucla methodi analyticae inventionem perperam Platoni tribuere. Bruckerum rectius scripsisse existimo; scilicet eos, qui Platonem hanc methodum invenisse volunt, non cogitare, illum audivisse Theodorum Cyrenaeum, celeberrimum Geometram, quem hanc rationem reducendi quaestiones ad sua principia ignoravisse, non verosimile est (Bruckeri, *Hist. Crit. Phil.*, tom. I. p. 642)'—De Gelder, *Theonis Smyrn. Arithm.*, Praemonenda, p. xlix., Lugd. Bat.

[55] Xenophon, *Memorab.*, IV. 7; Diog. Laert., II. 32, p. 41, ed. Cobet.

to learn geometry from Theodorus, and then to the Pythagoreans in Italy. Is it likely, then, that Plato, who, as far as we know, never solved a geometrical question, should have invented this method of solving problems in geometry and taught it to Archytas, who was probably his teacher, and who certainly was the foremost geometer of that time, and that thereby Archytas was led to his celebrated solution of the Delian problem?

The former of the two reasons advanced by Bretschneider, and given above, has reference to and is based upon the following well-known and remarkable passage of the *Republic* of Plato. The question under consideration is the order in which the sciences should be studied: having placed arithmetic first, and geometry—*i. e.* the geometry of plane surfaces—second, and having proposed to make astronomy the third, he stops and proceeds:—

" 'Then take a step backward, for we have gone wrong in the order of the sciences.'

'What was the mistake?' he said.

'After plane geometry,' I said, 'we took solids in revolution, instead of taking solids in themselves; whereas, after the second dimension the third, which is concerned with cubes and dimensions of depth, ought to have followed.'

'That is true, Socrates; but these subjects seem to be as yet hardly explored.

'Why, yes,' I said, 'and for two reasons: in the first place, no government patronises them, which leads to a want of energy in the study of them, and they are difficult; in the second place, students cannot learn them unless they have a teacher. But then a teacher is hardly to be found; and even if one could be found, as matters now stand, the students of these subjects, who are very conceited, would not mind him. That, however, would be otherwise if the whole state patronised and honoured this science; then they would listen, and there would be continuous and earnest search, and discoveries would be made; since even now, disregarded as these studies are by the world, and maimed of their fair proportions, and although none of their votaries can tell the use of them, still they force their way by their natural charm, and very likely they may emerge into light.'

'Yes,' he said, 'there is a remarkable charm in them. But I do not clearly understand the change in the order. First you began with a geometry of plane surfaces?'

'Yes,' I said.

'And you placed astronomy next, and then you made a step backward?'

'Yes,' I said, 'the more haste the less speed; the ludicrous state of solid geometry made me pass over this branch and go on to astronomy, or motion of solids.'

'True,' he said.

'Then regarding the science now omitted as supplied, if only encouraged by the State, let us go on to astronomy.'

'That is the natural order,' he said."[56]

Cantor, too, says that 'stereometry proper, notwithstanding the knowledge of the regular solids, seems on the whole to have been yet [at the time of Plato] in a very backward state,'[57] and in confirmation of his opinion quotes part of a passage from the *Laws*.[58] It will be seen, however, on reading it to the end, that the ignorance of the Hellenes referred to by Plato, and denounced by him in such strong language, is an ignorance—not, as Cantor thinks, of stereometry—but of incommensurables.

We do not know the date of the *Republic*, nor that of the discovery of the cubature of the pyramid by Eudoxus,

[56] Plato, *Rep.*, VII. 528; Jowett, *the Dialogues of Plato*, vol. II., pp. 363, 364.

[57] Cantor, *Gesch. der Math.*, p. 193.

[58] Plato, *Leges*, VII., 819, 820; Jowett, *op. cit.*, vol. IV., pp, 333, 334.

which founded stereometry,[59] and which was an important advance in the direction indicated in the passage given above: it is probable, however, that Plato had heard from his Pythagorean teachers of this desideratum; and I have in the last chapter (p. 86, *sq.*) pointed out a problem of high philosophical importance to the Pythagoreans at that time, which required for its solution a knowledge of stereometry. Further, the investigation given above shows, as Cantor remarks, that Archytas formed an honourable exception to the general ignorance of geometry of three dimensions complained of by Plato. It is noteworthy that this difficult problem—the cubature of the pyramid—was solved, not through the encouragement of any State, as suggested by Plato, but, and in Plato's own lifetime, by a solitary thinker—the great man whose important services to geometry we have now to consider.

[59] It should be noticed, however, that with the Greeks *Stereometry* had the wider signification of geometry of three dimensions, as may be seen from the following passage in Proclus: *ἡ μὲν γεωμετρία διαιρεῖται πάλιν εἴς τε τὴν ἐπίπεδον θεωρίαν καὶ τὴν στερεομετρίαν.*—Proclus, ed. Friedlein, p. 39: see also *ibid.*, pp. 73, 116.

CHAPTER V.

EUDOXUS.

Eudoxus of Cnidus.—His life.—Founded the School of Cyzicus.—Notices of his Geometrical work.—Examination of and inferences from these Notices.—Eudoxus discovered the Cubature of the Pyramid, invented the Method of Exhaustions, and was the Founder of the Doctrine of Proportion as given in the Fifth Book of Euclid.—καμπύλαι γραμμαὶ and Hippopede.—Retrospective view of the progress of Geometry.—Effect of the Dialectic Method in general, and, in particular, in Geometry.—Necessity of recasting the methods of investigation and proof.—Estimate of the services of Eudoxus.—Though his fame was very great in antiquity, yet he was for centuries unduly depreciated.—Justice is now done to him.—His place in the History of Science.

EUDOXUS of Cnidus[1]—astronomer, geometer, physician, lawgiver—was born about 407 B.C., and was a pupil of Archytas in geometry, and of Philistion, the Sicilian [or Italian Locrian], in medicine, as Callimachus relates in his *Tablets*. Sotion in his *Successions*, moreover, says that he also heard Plato; for when he was twenty-three years of age and in narrow circumstances, he was attracted by the reputation of the Socratic school, and, in company with Theomedon the physician, by whom he was supported, he went to Athens, where—or rather at Piræus—he remained two months, going each day to the city to hear the lectures of the Sophists, Plato being one of them, by whom, however, he was coldly received. He then returned home, and, being again aided by the contributions of his friends, he set sail for Egypt with Chrysippus—also a physician, and who, as well as Eudoxus, learnt medicine from Philistion—bearing with him letters of recommendation from Agesilaus to Nectanabis, by whom he was commended to

[1] Diog. Laert., VIII., c. viii; A. Boeckh, *ueber die vierjährigen Sonnenkreise der Alten, vorzüglich den Eudoxischen*, Berlin, 1863.

the priests. When he was in Egypt with Chonuphis of Heliopolis, Apis licked his garment, whereupon the priests said that he would be illustrious (ἔνδοξον), but short-lived.[2] He remained in Egypt one year and four months, and composed the *Octaëteris*[3]—an octennial period. Eudoxus then—his years of study and travel now over—took up his abode at Cyzicus, where he founded a school (which became famous in geometry and astronomy), teaching there and in the neighbouring cities of the Propontis; he also went to Mausolus. Subsequently, at the height of his reputation, he returned to Athens, accompanied by a great many pupils, for the sake, as some say, of annoying Plato, because formerly he had not held him worthy of attention. Some say that, on one occasion, when Plato gave an entertainment, Eudoxus, as there were many guests, introduced the fashion of sitting in a semicircle.[4] Aristotle tells us that Eudoxus thought that pleasure was the *summum bonum;* and, though dissenting from his theory, he praises Eudoxus in a manner which with him is quite unusual:—'And his words were believed, more from the excellence of his character than for themselves; for he had the reputation of being singularly virtuous, σώφρων: it therefore seemed that he did not hold this language as being a

[2] Boeckh thinks, and advances weighty reasons for his opinion, that the voyage of Eudoxus to Egypt took place when he was still young—that is, about 378 B. C.; and not in 362 B. C., in which year it is placed by Letronne and others. Boeckh shows that it is probable that the letters of recommendation from Agesilaus to Nectanabis, which Eudoxus took with him, were of a much earlier date than the military expedition of Agesilaus to Egypt. In this view Grote agrees. See Boeckh, *Sonnenkreise*, pp. 140–148; Grote, *Plato*, vol. I., pp. 120–124.

[3] The Octaëteris was an intercalary cycle of eight years, which was formed with the object of establishing a correspondence between the revolutions of the sun and moon; eight lunar years of 354 days, together with three months of 30 days each, make up 2922 days: this is precisely the number of days in eight years of 365¼ days each. This period, therefore, presupposes a knowledge of the true length of the solar year; its invention, however, is attributed by Censorinus to Cleostratus.

[4] Is this the foundation of the statement in Grote's *Plato*, vol. I., p. 124—'the two then became friends'?

friend to pleasure, but that the case really was so.'[5] On his return to his own country he was received with great honours—as is manifest, Diogenes Laertius adds, from the decree passed concerning him—and gave laws to his fellow-citizens; he also wrote treatises on astronomy and geometry, and some other important works. He was accounted most illustrious by the Greeks, and instead of Eudoxus they used to call him Endoxus, on account of the brilliancy of his fame. He died in the fifty-third year of his age, *circ.* 354 B. C.

The above account of the life of Eudoxus, with the exception of the reference to Aristotle, is handed down by Diogenes Laertius, and rests on good authorities.[6] Unfortunately, some circumstances in it are left undetermined as to the time of their occurrence. I have endeavoured to present the events in what seems to me to be their natural sequence. I regret, however, that in a few particulars as to their sequence I am obliged to differ from Boeckh, who has done so much to give a just view of the life and career of Eudoxus, and of the importance of his work, and of the high character of the school founded by him at Cyzicus. Boeckh thinks it likely that Eudoxus heard Archytas in geometry, and Philistion in medicine, in the interval between his Egyptian journey and his abode at Cyzicus.[7]

Grote, too, in the notice which he gives of Eudoxus, takes the same view. He says:—'Eudoxus was born in poor circumstances; but so marked was his early promise, that some of the medical school at Knidus assisted him to prosecute his studies—to visit Athens, and hear the Sophists, Plato among them—to visit Egypt, Tarentum

[5] Aristot. *Eth. Nic.*, x. 2, p. 1172, ed. Bek.

[6] Callimachus of Cyrene; he was invited by Ptolemy II., Philadelphus, to a place in the Museum; and was chief librarian of the library of Alexandria; he held this office from about 250 B. C. until his death, about 240 B. C. Hermippus of Smyrna. Sotion of Alexandria flourished at the close of the third century B. C. Apollodorus of Athens flourished about the year 143 B. C.—Smith's *Dictionary*.

[7] Boeckh, *Sonnenkreise*, p. 149.

(where he studied geometry with Archytas), and Sicily (where he studied τὰ ἰατρικὰ with Philistion). These facts depend upon the Πίνακες of Kallimachus, which are good authority (Diog. Laert. VIII. 86).'[8]

Now I think it is much more likely that, as narrated above, Eudoxus went in his youth from Cnidus to Tarentum—between which cities, as we have seen, an old commercial intercourse existed[9]—and there studied geometry under Archytas, and that he then studied medicine under the Sicilian [or Italian Locrian] Philistion. In support of this view, it is to be observed that—

1°. The narrative of Diogenes Laertius commences with this statement, which rests on Callimachus, who is good authority;

2°. The life of Eudoxus is given by Diogenes Laertius in his eighth book, which is devoted exclusively to the Pythagorean philosophers: this could scarcely have been so, if he was over thirty years of age when he heard Archytas, and that, too, only casually, as some think;

3°. The statement that he went from Tarentum to Sicily [or the Italian Locri] to hear Philistion, who probably was a Pythagorean—for we know that medicine was cultivated by the Pythagoreans—is in itself credible;

4°. Chrysippus, the physician in whose company Eudoxus travelled to Egypt, was also a pupil of Philistion in medicine, and Theomedon, with whom Eudoxus went to Athens, was a physician likewise; in this way might arise the relation between Eudoxus and some of the medical school of Cnidus noticed by Grote.

The statement of Grote, that 'these facts depend on the Πίνακες of Kallimachus,' is not correct; nor is there any authority for his statement that Eudoxus was assisted by the medical school of Cnidus to visit Tarentum and Sicily:

[8] Grote, *Plato,* vol. I., p. 123, *n.*

[9] *Supra*, p. 19: Herod., III. 138.

the probability is that he became acquainted with some physicians of Cnidus as fellow-pupils of Philistion.

The geometrical works of Eudoxus have unfortunately been lost; and only the following brief notices of them have come down to us:—

(*a*). Eudoxus of Cnidus, a little younger than Leon, and a companion of Plato's pupils, in the first place increased the number of general theorems, added three proportions to the three already existing, and also developed further the things begun by Plato concerning the section [of a line], making use, for the purpose, of the analytical method;[10]

(*b*). The discovery of the three later proportions, referred to by Eudemus in the passage just quoted, is attributed by Iamblichus to Hippasus, Archytas, and Eudoxus;[11]

(*c*). Proclus tells us that Euclid collected the elements, and arranged much of what Eudoxus had discovered.[12]

(*d*). We learn further from an anonymous scholium on the Elements of Euclid, which Knoche attributes to Proclus, that the Fifth Book, which treats of proportion, is common to geometry, arithmetic, music, and, in a word, to all mathematical science; and that this Book is said to be the invention of Eudoxus (Εὐδόξου τινὸς τοῦ Πλάτωνος διδασκάλου);[13]

(*e*). Diogenes Laertius tells us, on the authority of the Chronicles of Apollodorus, that Eudoxus was the discoverer of the theory of curved lines (εὑρεῖν τε τὰ περὶ τὰς καμπύλας γραμμάς);[14]

(*f*). Eratosthenes says, in the passage quoted above

[10] Proclus, ed. Friedlein, p. 67: see Introduction, p. 4.

[11] Iambl. *in Nic. Arithm.*, ed. Tennulius, pp. 142, 159, 163.

[12] Proclus, ed. Friedlein, p. 68: see introduction, p. 5.

[13] Euclidis *Elem.*, ed. August., pars ii., p. 328; Knoche, *Untersuchungen*, &c., p. 10: see p. 49.

[14] Diog. Laert., VIII., c. viii., ed. Cobet, p. 226.

(p. 111), that Eudoxus employed these so called curved lines to solve the problem of finding two mean proportionals between two given lines;[15] and in the epigram which concludes his letter to Ptolemy III., Eratosthenes associates him with Archytas and Menaechmus;[16]

(*g*). In the history of the 'Delian Problem' given by Plutarch, Plato is stated to have referred the Delians, who implored his aid, to Eudoxus of Cnidus or to Helicon of Cyzicus, for its solution;[17]

(*h*). We learn from Seneca that Eudoxus first brought back with him from Egypt the knowledge of the motions of the planets;[18] and from Simplicius, on the authority of Eudemus, that, in order to explain these motions, and in particular the retrograde and stationary appearances of the planets, Eudoxus conceived a certain curve, which he called the *hippopede*;[19]

(*i*). Archimedes tells us expressly that Eudoxus discovered the following theorems:—

> Any pyramid is the third part of a prism which has the same base and the same altitude as the pyramid;
>
> Any cone is the third part of a cylinder which has the same base and the same altitude as the cone.[20]

(*j*). Archimedes, moreover, points out the way in which these theorems were discovered: he tells us that he himself obtained the quadrature of the parabola by means of the following lemma:—'If two spaces are unequal, it is possible to add their difference to itself so often that every

[15] Archim., ed. Torelli, p. 144; ed. Heiberg., III., p. 106.

[16] Archim., ed. Tor., p. 146; ed. Heib., III., p. 112. Some writers translate θεουδέος in this epigram by 'divine,' but the true sense seems to be 'God-fearing,' 'pious': see Arist. (p. 129, *supra*).

[17] Plutarch, *de Gen. Soc.* I, *Opera*, ed. Didot, vol. III., p. 699.

[18] Seneca, *Quaest. Nat.*, VII. 3.

[19] Brandis, *Scholia in Aristot.*, p. 500ᵃ.

[20] Archim., ed. Torelli, p. 64; ed. Heiberg, vol. I., p. 4.

finite space can be surpassed. Former geometers have also used this lemma; for, by making use of it, they proved that circles have to each other the duplicate ratio of their diameters, and that spheres have to each other the triplicate ratio of their diameters; further, that any pyramid is the third part of a prism which has the same base and the same altitude as the pyramid; and that any cone is the third part of a cylinder which has the same base and the same altitude as the cone.'[21]

Archimedes, moreover, enunciates the same lemma for lines and for volumes, as well as for surfaces.[22] And the fourth definition of the Fifth Book of Euclid—which Book, we have seen, has been ascribed to Eudoxus—is somewhat similar.[23] It should be observed that Archimedes does not say that the lemma used by former geometers was exactly the same as his, but like it: his words are:—*ὁμοῖον τῷ προειρημένῳ λῆμμά τι λαμβάνοντες ἔγραφον.*

Concerning the three new proportions referred to in (*a*) and (*b*), see pp. 44, 45. In Proclus they are ascribed to Eudoxus; whereas Iamblichus reports that they are the invention of Archytas and Hippasus, and says that Eudoxus and his school (*οἱ περὶ Εὔδοξον μαθηματικοὶ*) only changed their names. The explanation of these conflicting statements, as Bretschneider has suggested, probably lies in this—that Eudoxus, as pupil of Archytas, learned these proportions from his teacher, and first brought them to Greece, and that later writers then believed him to have been the inventor of them.[24]

[21] Archim., ed. Tor., p. 18; ed. Heib., vol. II., p. 296.

[22] *Ἔτι δὲ τῶν ἀνίσων γραμμῶν καὶ τῶν ἀνίσων ἐπιφανειῶν καὶ τῶν ἀνίσων στερεῶν τὸ μεῖζον τοῦ ἐλάσσονος ὑπερέχειν τοιούτῳ, ὃ συντιθέμενον αὐτὸ ἑαυτῷ δυνατόν ἐστιν ὑπερέχειν παντὸς τοῦ προτεθέντος τῶν πρὸς ἄλληλα λεγομένων.* Archim., ed. Tor., p. 65; ed. Heib., vol. I., p. 10.

[23] This definition is—

Λόγον ἔχειν πρὸς ἄλληλα μεγέθη λέγεται, ἃ δύναται πολλαπλασιαζόμενα ἀλλήλων ὑπερέχειν.

[24] Bretsch., *Geom. vor Eukl.*, p. 164.

For additional information on this subject, and with relation to the further development of this doctrine by later Greek mathematicians, who added four more means to the six existing at this period, the reader is referred to Pappus, Nicomachus, Iamblichus, and also to the observations of Cantor with relation to them.[25]

The passage (*a*) concerning the section (περὶ τὴν τομὴν) was for a long time regarded as extremely obscure: it was explained by Bretschneider as meaning the section of a straight line in extreme and mean ratio, *sectio aurea*, and in the *Introduction* (p. 4, note) I adopted this explanation. Bretschneider's interpretation has since been followed by Cantor in his classical work on the *History of Mathematics*,[26] and may now be regarded as generally accepted.

A proportion contains in general four terms: the second and third terms may, however, be equal, and then three magnitudes only are concerned: further, if the magnitudes are *lines*, the third term may be the difference between the first and second, and thus the geometrical and arithmetical ratios may occur in the same proportion: the greatest line is then the sum of the two others, and is said to be cut in extreme and mean ratio. The construction of the regular pentagon depends ultimately on this section—which Kepler says was called *sectio aurea*, *sectio divina*, and *proportio divina*, on account of its many wonderful properties. This problem, to cut a given straight line in extreme and mean ratio, is solved in Euclid II. 11, and VI. 30; and the solution depends on the application of areas, which Eudemus tells us was an invention of the Pythagoreans. Use is made of the problem in Euclid IV. 10–14; and the subject is again taken up in the Thirteenth Book of the Elements.

Bretschneider observes that the first five propositions

[25] Pappi *Collect.*, ed. Hultsch, vol. I., p. 70, *sq.*; Cantor, *Gesch. der Math.* p. 206.

[26] *Ibid.*, p 208.

ment will be cut in a similar manner by taking on it a part equal to the less; and so on continually; and it results from Prop. III. that twice the lesser segment exceeds the greater. If now reference be made to the Tenth Book, which treats of incommensurable magnitudes, we find that the first proposition is as follows:—'Two unequal magnitudes being given, if from the greater a part be taken away which is greater than its half, and if from the remainder a greater part than its half, and so on, there will remain a certain magnitude which will be less than the lesser given magnitude'; and that the second proposition is—'Two unequal magnitudes being proposed, if the lesser be continually taken away from the greater, and if the remainder never measures the preceding remainder, these magnitudes will be incommensurable'; lastly, in the third proposition we have the method of finding the greatest common measure of two given commensurable magnitudes. Taking these propositions together, and considering them in connection with those in the Thirteenth Book, referred to above, it seems likely that the writer to whom the early propositions of the Tenth Book are due had in view the section of a line in extreme and mean ratio, out of which problem I have expressed the opinion that the discovery of incommensurable magnitudes arose (see p. 42).

This, I think, affords an explanation of the place occupied by Eucl. X. 1 in the Elements, which would otherwise be difficult to account for: we might rather expect to find it at the head of Book XII., since it is the theorem on which the Method of Exhaustions, as given by Euclid in that book, is based, and by means of which the following theorems in it are proved:—

Circles are to each other as the squares on their diameters, XII. 2:

A pyramid is the third part of a prism having the same base and same height, XII. 7;

A cone is the third part of a cylinder having the same base and same height, XII. 10;

Spheres are to each other in the triplicate ratio of their diameters, XII. 18.

Now two of the foregoing theorems are attributed to Eudoxus by Archimedes; and the lemma, which Archimedes tells us former geometers used in order to prove these theorems, is substantially the same as that assumed by Euclid in the proof of the first proposition of his Tenth Book: it is probable, therefore, that this proposition also is due to Eudoxus.

Eudoxus, therefore, as I have said (p. 96), must be regarded as the inventor of the Method of Exhaustions. We know, too, that the doctrine of proportion, as contained in the Fifth Book of Euclid, is attributed to him. I have, moreover, said (*loc. cit.*) that 'the invention of rigorous proofs for theorems such as Euclid VI. 1, involves, in the case of incommensurable quantities, the same difficulty which is met with in proving rigorously the four theorems stated by Archimedes in connection with this axiom.'[31] In all these cases the difficulty was got over, and rigorous proofs supplied, in the same way—namely, by showing that every supposition contrary to the existence of the properties in question led, of necessity, to some contradiction, in short by the *reductio ad absurdum*[32] (ἀπαγωγὴ εἰς ἀδύνατον). Hence it follows that Eudoxus must have been

[31] 'C'était encore par la réduction à l'absurde que les anciens étendaient aux quantités incommensurables les rapports qu'ils avaient découverts entre les quantités commensurables' (Carnot, *Réflexions sur la Métaphysique du Calcul Infinitésimal*, p. 137, second edition: Paris, 1813).

If the bases of the triangles are commensurable, this theorem, Euclid VI. 1, can be proved by means of the First Book and the Seventh Book, which latter contains the theory of proportion for numbers and for commensurable magnitudes. It is easy to see, then, that this theorem can be proved in a general manner—so as to include the case where the bases are incommensurable—by the method of *reductio ad absurdum* by means of the axiom used in Euclid X. 1, which has been attributed above to Eudoxus: see p. 134.

[32] Carnot, *ibid.*, p. 135.

familiar with this method of reasoning. Now this indirect kind of proof is merely a case of the Analytical Method, and is indeed the case in which the subsequent synthesis, that is usually required as a complement, may be dispensed with. In connection with this it may be observed that the term used here, ἀπαγωγή, is the same that we met with (p. 41, *n.* 62) on our first introduction to the analytical method; this indeed is natural, for analysis, as Duhamel remarked, is nothing else but a method of reduction.[33]

Eutocius, in his Commentary on the treatise of Archimedes *On the Sphere and Cylinder*, in which he has handed down the letter of Eratosthenes to Ptolemy III., and in which he has also preserved the solutions of the Delian Problem by Archytas, Menaechmus, and other eminent mathematicians, with respect to the solution of Eudoxus, merely says:

'We have met with the writings of many illustrious men, in which the solution of this problem is professed; we have declined, however, to report that of Eudoxus, since he says in the introduction that he has found it by means of curved lines, καμπύλων γραμμῶν: in the proof, however, he not only does not make any use of these curved lines, but also, finding a discreet proportion, takes it as a continuons one; which was an absurd thing to conceive—not merely for Eudoxus, but for those who had to do with geometry in a very ordinary way.'[34]

As Eutocius omitted to transmit the solution of Eudoxus, so I did not give the above with the other notices of his geometrical work. It is quite unnecessary to defend Eudoxus from either of the charges contained in this passage. I will only remark, with Bretschneider, that it

[33] 'L'analyse n'est donc autre chose qu'une méthode de *réduction*' (Duhamel, *des Méthodes dans les Sciences de Raisonnement*, première partie, p. 41).

[34] Archim., ed. Tor., p. 135; ed. Heiberg, vol. VI. p. 66.

is strange that Eutocius, who had before him the letter of Eratosthenes, did not recognise in the complete corruption of the text the source of the defects which he blames.[35]

We have no further notice of these so called curved lines: it is evident, however, that they could not have been any of the conic sections, which were only discovered later by Menaechmus, the pupil of Eudoxus.

There is a conjecture, however, concerning them which is worth noticing: M. Paul Tannery thinks that the term καμπύλαι γραμμαὶ has, in the text of Eratosthenes, a particular signification, and that, compared with, *e. g.* the καμπύλα τόξα of Homer, it suggests the idea of a curve symmetrical to an axis, which it cuts at right angles, and presenting an inflexion on each side of this axis. Tannery conjectures that these curves of Eudoxus are to be found amongst the projections of the curves used in the solution of his master, Archytas; and tries to find whether, amongst these projections, any can be found to which the denomination in question can be suitably applied. We have seen above, pp. 119, 120, that Flauti has shown how the solution of Archytas could be constructed by means of the projections, on one of the vertical planes, of the curves employed in that solution. I have further shown that the actual construction of these projections can be obtained by the aid of geometrical theorems and problems known at the time of Archytas; though we have no evidence that he completed his solution in this way. Tannery has considered these curves, and shown that the term καμπύλαι γραμμαὶ, in the sense which he attaches to it, does not apply to either of them, nor to the projections on the other vertical plane; but that, on the contrary, the term is quite applicable to the projection of the intesrection of the cone and tore on the circular base of the cylinder.[36]

35 Bretsch., *Geom. vor Eukl.*, p. 166.

36 Tannery, *sur les Solutions du Problème de Délos par Archytas et par Eudoxe.*

The astronomical work of Eudoxus is beyond the scope of this treatise, and is only referred to in connection with the *hippopede* (*h*). I may briefly state, however, that he was a practical observer, and that he 'may be considered as the father of scientific astronomical observation in Greece'; further, that 'he was the first Greek astronomer who devised a systematic theory for explaining the periodic motions of the planets';[37] that he did so by means of geometrical hypotheses, which latter were submitted to the test of observations, and corrected thereby; and that hence arose the system of concentric spheres which made the name of Eudoxus so illustrious amongst the ancients.

Although this theory was substantially geometrical, and is in the highest degree worthy of the attention of the students of the history of geometry, yet to render an account of it which would be in the least degree satisfactory would altogether exceed the limits prescribed to me; I must, therefore, refer my readers to the excellent and memorable monograph of Schiaparelli,[38] who with great ability and with rare felicity has restored the work of Eudoxus. In this memoir the nature of the spherical curve, called by Eudoxus the *hippopede*, was first placed in a clear light: it is the intersection of a sphere and cylinder; and on account of its form, which resembles the figure

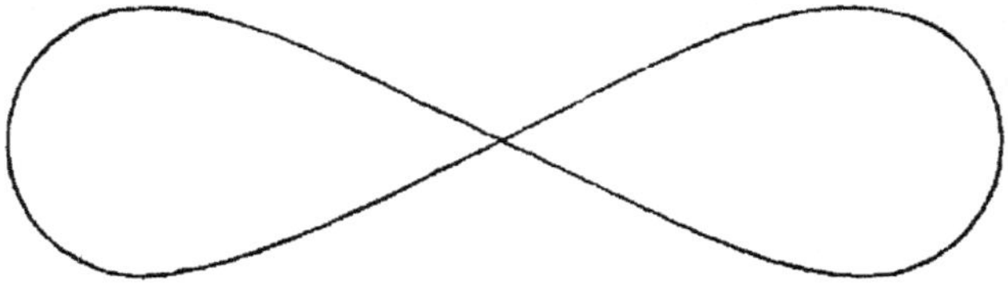

it is called by Schiaparelli a *spherical lemniscate*.[39] A passage in Xenophon, *de re equestri*, cap. 7, explains why

[37] Sir George Cornewall Lewis, *a Historical Survey of the Astronomy of the Ancients*, p. 147, *sq.*: London, 1862.

[38] G. V. Schiaparelli, *le Sfere Omocentriche di Eudosso, di Callippo e di Aristotele* (Ulrico Hoepli: Milano, 1875).

[39] See Schiaparelli, *loc. cit.*, section v.

the name *hippopede* was given to this curve, and also to one of the *spirics* (ἡ ἱπποπέδη, μία τῶν σπειρικῶν οὖσα)[40] of Perseus, which also has the form of a lemniscate.[41]

I have examined the work of Eudoxus, and pointed out the important theorems discovered by him; I have also dwelt on the importance of the methods of inquiry and proof which he introduced. In order to appreciate this part of his work, it seems desirable to take a brief retrospective glance at the progress of geometry as set forth in the former chapters of this work, and the state in which it was at the time of Eudoxus, and also to refer to the philosophical movement during the last generation of the fifth century, B. C. :—

In the first chapter (p. 14) I attributed to Thales the theorem that the sides of equiangular triangles are proportional; a theorem which contains the beginnings of the doctrine of proportion and of the similarity of figures. It is agreed on all hands that these two theories were treated at length by Pythagoras and his School. It is almost certain, however, that the theorems arrived at were proved for commensurable magnitudes only, and were assumed to hold good for all. We have seen, moreover, that the discovery of incommensurable magnitudes is attributed to Pythagoras himself by Eudemus: this discovery, and the construction of the regular pentagon, which involves incommensurability, depending as it does on the section of a line in extreme and mean ratio, were always regarded as glories of the School, and kept secret; and it is remarkable that the same evil fate is said to have overtaken the person who divulged each of these secrets—secrets, too, regarded by the brotherhood as so peculiar that the pen-

40 Proclus, ed. Friedlein, p. 127. With respect to the spiric lines, see Knoche and Maerker, *ex Procli successoris in Euclidis Elementa commentariis definitionis quartae expositionem quae de recta est linea et sectionibus spiricis commentati sunt J. H. Knochius et F. J. Maerkerus*, Herfordiae, 1856.

41 See *infra*, p. 156, note 9.

tagram, which might be taken to represent both these discoveries, was used by them as a sign of recognition. It seems to be a fair inference from what precedes, that the Pythagoreans themselves were aware that their proofs were not rigorous, and were open to serious objection;[42] indeed, after the invention of dialectics by Zeno, and the great effect produced throughout Hellas by his novel and remarkable negative argumentation, any other supposition is not tenable. Further, it is probable that the early Pythagoreans, who were naturally intent on enlarging the boundaries of geometry, took for granted as self-evident many theorems, especially the converses of those already established. The first publication of the Pythagorean doctrines was made by Philolaus; and Democritus, who was intimate with him, and probably his pupil, wrote on incommensurables.

Meanwhile the dialectic method and the negative mode of reasoning had become more general, or to use the words of Grote :—

'We thus see that along with the methodised question and answer, or dialectic method, employed from henceforward more and more in philosophical inquiries, comes out at the same time the negative tendency—the probing, testing, and scrutinising force—of Grecian speculation. The negative side of Grecian speculation stands quite as prominently marked, and occupies as large a measure of the intellectual force of their philosophers, as the positive side. It is not simply to arrive at a conclusion, sustained by a certain measure of plausible premise—and then to proclaim it as an authoritative dogma, silencing or disparag-

[42] A similar view of the subject is taken by P. Tannery, *de la solution géométrique des problèmes du second degré avant Euclide.* Mémoires de la Société des Sciences physiques et naturelles de Bourdeaux, tom. IV. (2e serie), p. 406. He says :—'La découverte de l'incommensurabilité de certaines longueurs entre elles, et avant tout de la diagonale du carré à son côté, qu'elle soit due au Maître ou aux disciples, dut, dès lors, être un véritable scandale logique, une redoutable pierre d'achoppement.'

ing all objectors—that Grecian speculation aspires. To unmask not only positive falsehood, but even affirmation without evidence, exaggerated confidence in what was only doubtful, and show of knowledge without the reality—to look at a problem on all sides, and set forth all the difficulties attending its solution—to take account of deductions from the affirmative evidence, even in the case of conclusions accepted as true upon the balance—all this will be found pervading the march of their greatest thinkers. As a condition of all progressive philosophy, it is not less essential that the grounds of negation should be freely exposed than the grounds of affirmation. We shall find the two going hand in hand, and the negative vein, indeed, the more impressive and characteristic of the two, from Zeno downward, in our history.'[43]

As an immediate consequence of this, it would follow that the truth of many theorems, which had been taken for granted as self evident, must have been questioned; and that, in particular, doubt must have been thrown on the whole theory of the similarity of figures and on all geometrical truths resting on the doctrine of proportion: indeed it might even have been asked what was the meaning of ratio as applied to incommensurables, inasmuch as their mere existence renders the arithmetical theory of proportion inexact in its very definition.[44]

Now it is remarkable that the doctrine of proportion is *twice* treated in the Elements—first, in a general manner, so as to include incommensurables, in Book V., which tradition ascribes to Eudoxus, and then arithmetically in Book VII., which probably, as Hankel has supposed, contains the treatment of the subject by the older Pythagoreans.[45] The twenty-first definition of Book VII. is—'Αριθμοὶ ἀνάλογόν εἰσιν, ὅταν ὁ πρῶτος τοῦ δευτέρου καὶ ὁ τρίτος τοῦ

[43] Grote, *History of Greece*, vol. VI. p. 48.

[44] See the Articles on *Proportion* and *Ratio* in the English Cyclopaedia.

[45] Hankel, *Gesch. der Math.*, p. 390.

τετάρτου ἰσάκις ᾖ πολλαπλάσιος, ἢ τὸ αὐτὸ μέρος, ἢ τὰ αὐτὰ μέρη ὦσιν.

Further, if we compare this definition with the third, fourth, and fifth definitions of Book V., I think we can see evidence of a gradual change in the idea of ratio, and of a development of the doctrine of proportion—

1. The third definition, which is generally considered not to belong to Euclid,[46] seems to be an attempt to bridge over the difficulty which is inherent in incommensurables, and may be a survival of the manner in which the subject was treated by Democritus.

2. The fourth definition[47] is generally regarded as having for its object the exclusion of the ratios of finite magnitudes to magnitudes which are infinitely great on the one side, and infinitely small on the other: it seems to me, however, that its object may have been, rather, to include the ratios of incommensurable magnitudes: moreover, since the doctrine of proportion by means of the apagogic method of proof can be founded on the axiom which is connected with this definition, and which is the basis of the *method of exhaustions*, it is possible that the subject may have been first presented in this manner by Eudoxus.

3. Lastly, in the fifth definition his final and systematic manner of treating the subject is given.[48]

Those who are acquainted with the history of Greek philosophy know that a state of things somewhat similar to

[46] Λόγος ἐστὶ δύο μεγεθῶν ὁμογενῶν ἡ κατὰ πηλικότητα πρὸς ἄλληλα ποιὰ σχέσις. See Camerer, *Euclidis Elementorum libri sex priores*, tom. II. p. 74, *sq.*, Berolini, 1824.

[47] Λόγον ἔχειν πρὸς ἄλληλα μεγέθη λέγεται, ἃ δύναται πολλαπλασιαζόμενα ἀλλήλων ὑπερέχειν.

[48] In connection with what precedes, we are reminded of the aphorism of Aristotle—'We cannot prove anything by starting from a different genus, *e.g.* nothing geometrical by means of arithmetic. . . . . Where the subjects are so different as they are in arithmetic and geometry we cannot apply the arithmetical sort of proof to that which belongs to quantities in general, unless these quantities are numbers, which can only happen in certain cases.' *Anal. post.* I. vii. p. 75[a], ed. Bek.

that represented above existed with respect to it also, and that a problem of a similar character, also requiring a new method, proposed itself for solution towards the close of the fifth century B.C.; and, further, that this problem was solved by Socrates by means of a new philosophic method—the analysis of general conceptions. This must have been known to Eudoxus, for we are informed that he was attracted to Athens by the fame of the Socratic School. Now a service, similar to that rendered by Socrates to philosophy, but of higher importance, was rendered by Eudoxus to geometry, who not only completed it by the foundation of stereometry, but, by the introduction of new methods of investigation and proof, placed it on the firm basis which it has maintained ever since.

This eminent thinker—one of the most illustrious men of his age, an age so fruitful in great men, the precursor, too, of Archimedes and of Hipparchus—after having been highly estimated in antiquity,[49] was for centuries unduly depreciated;[50] and it is only within recent years that, owing to the labours of some conscientious and learned men, justice has been done to his memory, and his reputation restored to its original lustre.[51]

[49] *E.g.* Cicero, *de Div.* II. 42, 'Ad Chaldaeorum monstra veniamus: de quibus Eudoxus, Platonis auditor, in astrologia judicio doctissimorum hominum facile princeps, sic opinatur, id quod scriptum reliquit: Chaldaeis in praedictione et in notatione cujusque vitae ex natali die, minime esse credendum': Plutarch, *non posse suav. vivi sec. Epic.* c. xi. **Εὐδόξῳ δὲ καὶ Ἀρχιμήδει καὶ Ἱππάρχῳ συνενθουσιῶμεν.**

[50] As evidence of this depreciation I may notice—Delambre, *Histoire de l'Astronomie ancienne*, 'L'Astronomie n'a été cultivée véritablement qu'en Grèce, et presque uniquement par deux hommes, Hipparque et Ptolemée' (tom. I p. 325): 'Rien ne prouve qu'il [Eudoxe] fut géomètre' (tom. I. p. 131). Well may Schiaparelli say—'Questa enorme proposizione.' Equally monstrous is the following:—'It is only in the first capacity [astronomer and not geometer] that his fame has descended to our day, and he has more of it than can be justified by any account of his astronomical science now in existence.' De Morgan, in Smith's *Dictionary.*

[51] Ideler, *ueber Eudoxus*, Abh. der Berl. Akad. v. J. 1828 and 1830: Letronne, *sur les écrits et les travaux d'Eudoxe de Cnide, d'après M. Ludwig*

Something, however, remained to be cleared up, especially with regard to his relations, and supposed obligations, to Plato.[52] I am convinced that the obligations were quite in the opposite direction, and that Plato received from Eudoxus incomparably more than he gave. As to his solving problems proposed by Plato, the probablity is that these questions were derived from the same source—Archytas and the Pythagoreans. Yet I attach the highest importance to the visit of Eudoxus to Athens; for although he heard Plato for two months only, that time was sufficient to enable Eudoxus to become acquainted with the Socratic method, to see that it was indispensable to clear up some of the fundamental conceptions of geometry, and, above all, to free astronomy from metaphysical mystifications, and to render the treatment of that science as real and positive as that of geometry. To accomplish this, however, it was incumbent on him to know the celestial phenomena, and for this purpose—inasmuch as one human life was too short—he saw the necessity of going to Egypt, to learn from the priests the facts which an observation continued during many centuries had brought to light, and which were there preserved.

I would call particular attention to the place which Eudoxus filled in the history of science—with him, in fact, an epoch closed, and a new era, still in existence, opened.[53]

Ideler, Journal des Savants, 1840: Boeckh, *Sonnenkreise der Alten*, 1863: Schiaparelli, *le Sfere Omocentriche*, &c., 1875.

[52] Even those by whom the fame of Eudoxus has been revived seem to acquiesce in this.

[53] This has been pointed out by Auguste Comte:—'Celle-ci [la seconde évolution scientifique de la Grèce] commença pourtant, avec tous ses caractères propres, pendant la génération anterieure à cette ère [la fondation du Musée d'Alexandrie], chez un savant trop méconnu, qui fournit une transition normale entre ces deux grandes phases théoriques, composées chacune d'environ trois siècles. Quoique nullement philosophe, Eudoxe de Cnide fut le dernier théoricien embrassant, avec un égal succès, toutes les spéculations accessibles à l'ésprit mathématique. Il servit pareillement la géométrie et l'astronomie, tandis que, bientôt après lui, la spécialisation devint déjà telle que ces deux sciences ne purent plus être notablement perfectionnées par les mêmes organes.' *Politique Positive*, III., p. 316, Paris, 1853.

He was geometer, astronomer, physician, lawgiver, and was also counted amongst the Pythagoreans, and versed in the philosophy of his time. He was, however, much more *the man of science*, and, of all the ancients, no one was more imbued with the true scientific and positive spirit than was Eudoxus: in evidence of this, I would point to—

1°. His work in all branches of the geometry of the day—founding new, placing old on a rational basis, and throwing light on all—as presented above.

2°. The fact that he was the first who made observation the foundation of the study of the heavens, and thus became the father of true astronomical science.

3°. His geometrical hypothesis of concentric spheres, which was conceived in the true scientific spirit, and which satisfied all the conditions of a scientific research, even according to the strict notions attached to that expression at the present day.

4°. His 'practical and positive genius, which was averse to all idle speculations.'[54]

5°. The purely scientific school founded by him at Cyzicus, and the able mathematicians who issued from that school, and who held the highest rank as geometers and astronomers in the fourth century B. C.

We see, then, in Eudoxus something quite new—the first appearance in the history of the world of the man of science; and, as in all like cases, this change was

[54] Ideler, and after him Schiaparelli: this appears from the fact testified by Cicero (*vid. supra, n.* 49), that Eudoxus had no faith in the Chaldean astrology which was then coming into fashion among the Greeks; and also from this—that he did not, like many of his predecessors and contemporaries, give expression to opinions upon things which were inaccessible to the observations and experience of the time. An instance of this is found in Plutarch (*non posse suav. viv. sec. Epic.* CXI., vol. iv., p. 1138, ed. Didot), who relates that he, instead of speculating, as others did, on the nature of the sun, contented himself with saying that 'he would willingly undergo the fate of Phaeton if, by so doing, he could ascertain its nature, magnitude, and form.'

effected by a man who was thoroughly versed in the old system.[55]

It is not without significance, too, that Eudoxus selected the retired and pleasant shores of the Propontis as the situation of the school which he founded for the transmission of his method. Among the first who arose in this school was Menaechmus, whose work I have next to consider.

It is pleasing to see that the number of students of the history of mathematics is ever increasing; and that the centres in which the subject is cultivated are becoming more numerous; it is particularly gratifying to observe that the subject has at last attracted attention in England. Dr. Heiberg, of Copenhagen, has completed his edition of Archimedes: Archimedis *Opera omnia cum commentariis Eutocii:* e codice Florentino recensuit, Latine vertit notisque illustravit J. L. Heiberg, Dr. Phil., vols. II. et III.: Lipsiae, 1881. Dr. Heiberg has been since engaged in bringing out, in conjunction with Professor H. Menge, a complete edition of the works of Euclid, of which two volumes have been published: Euclidis *Elementa*, edidit et Latine interpretatus est J. L. Heiberg, Dr. Phil., vol. I., Libros I–IV. continens, vol. II., Libros V.–IX continens, Lipsiae, 1883, 1884. As Heiberg's edition of Archimedes was preceded by his *Quaestiones Archimedeae*, Hauniae, 1879; so, in anticipation of his edition of Euclid he has published: *Litterargeschichtliche Studien über Euklid*, Leipzig, 1882, a valuable work, to which I have referred in the fourth chapter. Dr Hultsch, of Dresden, informs me that his edition of Autolycus is finished, and that he hopes it will appear at the end of this month (June, 1885). The publication of this work—in itself so important, inasmuch as the Greek text of the propositions only of Autolycus has been hitherto published—will have, moreover, an especial interest with regard to the subject of the pre-Euclidian geometry. The Cambridge Press announce a work by Mr. T. L. Heath (author of the Articles on 'Pappus' and 'Porisms' in the *Encyclopædia Britannica*) on Diophantus; a subject on which M. Paul Tannery also has been occupied for some time.

---

[55] Eudoxus may even be regarded as in a peculiar manner uniting in himself and representing the previous philosophic and scientific movement; for—though not an *Ionian*—he was a native of one of the neighbouring Dorian cities; he studied under the *Pythagoreans* in Italy; and, subsequently, he went to Athens, being attracted by the reputation of the *Socratic* school.

The following works on the history of Mathematics have been recently published :—

Marie, Maximilien, *Histoire des Sciences Mathématiques et Physiques*, tomes I.–V., Paris, 1883, 1884. The first volume alone—*de Thalès à Diophante*—treats of the subject of these Papers. It is, in my judgment, inferior to the *Histoire des Mathématiques* of M. Hoefer, notwithstanding the errors of the latter, to which I called attention in p. 2, note. For the historical part of this volume M. Marie has followed Montucla without making use, or even seeming to suspect the existence, of the copious and valuable materials which have of late years accumulated on this subject. Referring to this, Heiberg (*Philologus* XLIII. *Jahresberichte*, p. 324) says: 'The author has been engaged with his book for forty years: one would have thought rather that the book was written forty years ago.' M. Marie commences his *Preface* by saying: 'The history that I have desired to write is that of the filiation of ideas and of scientific methods;' as if that was not the aim of all recent enlightened inquiries. Hear what Hankel, in *Bullettino Boncompagni*, V. p. 300, says: *la Storia della matematica non deve semplicemente enumerare gli scienzati e i loro lavori, ma essa deve altresi esporre lo sviluppo interno delle idee che vegnano nella scienza* (Quoted by Heiberg in *Philologus*, l. c.).

Gow, James, *A Short History of Greek Mathematics*, Cambridge, 1884. This history, as far at least as geometry is concerned, is not, nor indeed does it pretend to be, a work of independent research. Unlike M. Marie, however, Mr. Gow has to some extent studied the recent works on the subject, and the reader will see that he has made much use of the early chapters of this work (published in HERMATHENA, No. V., 1877, and No. VII., 1881). On the other hand, he has left unnoticed many important publications. In particular, the numerous and valuable essays of M. Paul Tannery, which leave scarcely any department of ancient mathematics untouched, and which throw light on all, seem to be altogether unknown to him. Essays and monographs like those of M. Tannery and others are in fact, with the single exception of Cantor's *Vorlesungen über Geschichte der Mathematik*, the only works in which progress in the history of ancient mathematics has of late years been made: Bretschneider's *Geometrie vor Euklides* and Hankel's *Geschichte der Mathematik* are no exceptions; for the former work is a monograph, and the latter, which was interrupted by the death of the author, contains only some fragments of a history of mathematics, and consists in reality of a collection of essays. Should the reader look at Heiberg's Paper in the *Philologus*, XLIII., 1884, pp. 321–346, and pp. 467–522, which has been referred to above, he will see how numerous and how important are the publications on Greek mathematics which have appeared since the opening of a new period of mathematico-historical research with the works of Chasles and Nesselmann more than forty years ago.

A glance at the subjoined list of the Papers of a single writer—M. Paul

Tannery—relating to the period from Thales to Euclid, will enable the reader to form an opinion on the extent of the literature treated of by Dr. Heiberg.

Mémoires de la Société des Sciences physiques et naturelles de Bordeaux (2e Série).—Tome I, 1876, Note sur le système astronomique d'Eudoxe. Tome II., 1878, Hippocrate de Chio et la quadrature des lunules; Sur les solutions du problème de Délos par Archytas et par Eudoxe. Tome IV., 1882, De la solution géométrique des Problèmes du second degré avant Eudoxe. Tome V., 1883, Seconde note sur le système astronomique d'Eudoxe; Le fragment d'Eudème sur la quadrature des lunules.

Bulletin des Sciences Mathématiques et Astronomiques.—Tome VII., 1883, Notes pour l'histoire des lignes et surfaces courbes dans l'antiquité. Tome IX., 1885, Sur l'Arithmétique Pythagorienne. Le vrai problème de l'histoire des Mathématiques anciennes.

Annales de la faculté des lettres de Bordeaux.—Tome IV., 1882, Sur les fragments d'Eudème de Rhodes relatifs à l'histoire des mathématiques. Tome V., 1883, Un fragment de Speusippe.

Revue philosophique de France et de l'étranger, dirigée par M. Ribot.—Mars, 1880, Thalès et ses emprunts á Égypte.

Novembre, 1880, Mars, Août et Décembre, 1881, L'éducation Platonicienne.

CHAPTER VI.

THE SUCCESSORS OF EUDOXUS.

I. MENAECHMUS.

Notices of Menaechmus and of his work.—His Solution of the Problem of the Duplication of the Cube.—He discovered the three Conic Sections.—Passage from the 'Review of Mathematics' of Geminus quoted.—Hypothesis of Bretschneider as to the way in which Menaechmus was led to the discovery of the Conic Sections.—Comparison of these Investigations with the Solution of Archytas.—Various inferences from the notices of the work of Menaechmus considered.—Successors of Eudoxus in the School of Cyzicus.—Solution of the Problem of the Duplication of the Cube attributed to Plato.—Strong presumption against its being genuine.—Plato's Solution.—The Geometrical Theorems used in it were known to Archytas.—Recapitulation.

MENAECHMUS—pupil of Eudoxus, associate of Plato, and the discoverer of the conic sections—is rightly considered by Th. H. Martin[1] to be the same as the Manaechmus of Suidas and Eudocia, 'a Platonic philosopher of Alopeconnesus; but, according to some, of Proconnesus, who wrote philosophic works and three books on Plato's

[1] Theonis Smyrnaei Platonici *Liber de Astronomia*, Paris, 1849, p. 59. A. Boeckh (*ueber die vierjährigen Sonnenkreise der Alten*, Berlin, 1863, p. 152), Schiaparelli (*le Sfere Omocentriche di Eudosso, di Callippo e di Aristotele*, Milano, 1875, p. 7), and Zeller, (*Plato and the Older Academy*, p. 554, note (28), E. T.), hold the same opinion as Martin: Bretschneider (*Geom. vor Eukl.*, p. 162), however, though thinking it probable that they were the same, says that the question of their identity cannot be determined with certainty. Both Martin and Bretschneider identify Menaechmus Alopeconnesius with the one referred to by Theon in the fragment (*k*) given below. Max C. P. Schmidt (*Die Fragmente des Mathematikers Menaechmus*, Philologus, Band XLII., p. 77, 1884), on the other hand, holds that they were distinct persons, but says that it is certainly more probable that the Menaechmus referred to by Theon was the discoverer of the conic sections, than that he was the Alopeconnesian, inasmuch as Theon connects him with Callippus, and calls them both μαθηματικοί. Schmidt, however, does not give any reason in support of his opinion that the Alopeconnesian was a distinct person. But when we consider that Alopeconnesus was in the Thracian Chersonese, and not far from Cyzicus, and that Proconnesus, an island in the

Republic.' From the following anecdote, taken from the writings of the grammarian Serenus and handed down by Stobaeus, he appears to have been the mathematical teacher of Alexander the Great :—Alexander requested the geometer Menaechmus to teach him geometry concisely; but he replied: 'O king, through the country there are private and royal roads, but in geometry there is only one road for all."[2] We have seen that a similar story is told of Euclid and Ptolemy I. (p. 5).

What we know further of Menaechmus is contained in the following eleven fragments :[3]—

(*a*). Eudemus informs us in the passage quoted from Proclus in the Introduction (p. 4), that Amyclas of Heraclea, one of Plato's companions, and Menaechmus, a pupil of Eudoxus and also an associate of Plato, and his brother, Deinostratus, made the whole of geometry more perfect.[4]

(*b*.) Proclus mentions Menaechmus as having pointed out the two different senses in which the word element (στοιχεῖον) is used.[5]

(*c*.) In another passage Proclus, having shown that many so called conversions are false and are not properly

Propontis, was still nearer to Cyzicus, and that, further, the Menaechmus referred to in the extract (*k*) modified the system of concentric spheres of Eudoxus, the supposition of Th. M. Martin (*l. c.*) that this extract occurred in the work of the Alopeconnesian on Plato's *Republic* in connection with the distaff of the Fates in the tenth book becomes probable.

[2] Stobaeus, *Floril.*, ed. A. Meineke, vol. IV., p. 205. Bretschneider (*Geom. vor Eukl.*, p. 162) doubts the authenticity of this anecdote, and thinks that it may be only an imitation of the similar one concerning Euclid and Ptolemy. He does so on the ground that it is nowhere reported that Alexander had, besides Aristotle, Menaechmus as a special teacher in geometry. This is an insufficient reason for rejecting the anecdote, and, indeed, it seems to me that the probability lies in the other direction, for we shall see that Aristotle had direct relations with the school of Cyzicus.

[3] The fragments of Menaechmus have been collected and given in Greek by Max C. P. Schmidt (*l. c.*).

[4] Proclus, ed. Friedlein, p. 67.

[5] *Ibid.*, p. 72.

conversions, adds that this fact had not escaped the notice of Menaechmus and Amphinomus and the mathematicians who were their pupils.[6]

(*d.*) In a third passage of Proclus, where he discusses the division of mathematical propositions into problems and theorems, he says, that whilst in the view of Speusippus and Amphinomus and their followers all propositions were theorems, it was maintained on the contrary by Menaechmus and the mathematicians of his School (*οἱ περὶ Μέναιχμον μαθηματικοί*) that they should all be called problems—the difference being only in the nature of the question stated, the object being at one time to find the thing sought, at another time, taking a definite thing, to see either what it is, or of what kind it is, or what affection it has, or what relation it has to something else.[7]

(*e*). In a fourth passage Proclus mentions him as the discoverer of the conic sections. The passage is in many respects so interesting that it deserves to be quoted in full:

'Again, Geminus divides a line into the compound and the uncompounded—calling a compound that which is broken and forms an angle; then he divides a compound line into that which makes a figure, and that which may be produced *ad infinitum*, saying that some form a figure, *e. g.* the circle, the ellipse (*θυρεός*),[8] the cissoid, whilst others do not form a figure, *e. g.* the section of the right-angled cone

[6] Proclus, ed. Friedlein, pp. 253-4.

[7] *Ibid.*, pp. 77, 78.

[8] 'ὁ θυρεός (the door-shape, oblong: cf. Heron Alexandr., ed. Hultsch, *Definit.* 95, p. 27: ποιοῦσα σχῆμα θυροειδές). It is called by Eutocius, *Comm.* to Apollon., p. 10: ἔλλειψιν, ἣν καὶ θυρεὸν καλοῦσι, and is used several times in Proclus.' So Heiberg; who adds that in one passage it occurs in an extract from Eudemus, and says that we may perhaps assume that we have here the original name for the ellipse (*Nogle Puncter af de graeske Mathematikeres Terminologi*, Philologisk-historiske Sam funds Mindeskrift, Kjobenhavn, 1879, p. 7). With relation to the same term, Heiberg, in his *Studien über Euklid*, p. 88, quotes a passage of the Φαινόμενα of Euclid, which had hitherto been overlooked: ἐὰν γὰρ κῶνος ἢ κύλινδρος ἐπιπέδῳ τμηθῇ μὴ παρὰ τὴν βασιν, ἡ τομὴ γίγνεται ὀξυγωνίου κώνου τομή, ἥτις ἐστὶν ὁμοία θυρεῷ, ed. D. Gregory, p. 561; and says

[the parabola], the section of the obtuse-angled cone [the hyperbola], the conchoid, the straight line, and all such. And again, after another manner, of the uncompounded line one kind is simple and the other mixed; and of the simple, one forms a figure, as the circular; but the other is indefinite, as the straight line; but of the mixed, one sort is in planes, the other in solids; and of that in planes, one kind meets itself as the cissoid, another may be produced to infinity; but of that in solids, one may be considered in the sections of solids, and the other may be considered as [traced] around solids. For the helix, which is decribed about a sphere or cone, exists around solids, but the conic sections and the spirical are generated from such a section of solids. Further, as to these sections, the conics were conceived by Menaechmus, with reference to which Eratosthenes says—

> "Nor cut from a cone the Menaechmian triads";

and the latter [the spirics] were conceived by Perseus, who made an epigram on their invention:

> "Perseus found the three [spirical] lines in five sections,
> and in honour of the discovery sacrificed to the gods."

Now, on the one hand, the three sections of the cone are the parabola, the hyperbola, and the ellipse; and, on the other, of the spirical sections one kind is inwoven, like the *hippopede;*[9] and another kind is dilated in the middle, and

that θυρεός was probably the name by which the curve was known to Menaechmus. It may be observed, however, that an ellipse is not of the shape of a door, neither is a shield, which is a secondary signification of θυρεός; the primary signification of the word is not 'door,' but 'large stone' which might close the entrance to a cave, as in Homer (*Odyssey*, IX.): such a stone, or boulder, as may be met with on exposed beaches is often of a flattened oval form, and the names of a shield of such a shape, and of an ellipse, may have been thence derived.

[9] τῶν δὲ σπειρικῶν τομῶν ἡ μέν ἐστιν ἐμπεπλεγμένη, ἐοικυῖα τῇ τοῦ ἵππου πέδῃ. The *hippopede* is also referred to in the two following passages of Proclus ἡ ἱπποπέδη, μία τῶν σπειρικῶν οὖσα (ed. Fried., p. 127), and καίτοιγε ἡ κισσοειδὴς μία οὖσα ποιεῖ γωνίαν καὶ ἡ ἱπποπέδη (*ibid.*, p. 128). In p. 142 I said that a passage in Xenophon, *de re equestri*, cap. 7, explains why the name *hippopede* was given to the curve conceived by Eudoxus for the explanation of the motions of the

becomes narrow at each extremity; and another being oblong, has less distance in the middle, but is dilated on each side.'[10]

(*f*). The line from Eratosthenes, which occurs in the preceding passage, is taken from the epigram which closes his famous letter to Ptolemy III., and which has been already more than once referred to. We now cite it with its context :—

> *μηδὲ σύ γ' Ἀρχύτεω δυσμήχανα ἔργα κυλίνδρων*
> *μηδὲ Μενεχμείους κωνοτομεῖν τριάδας*
> *δίζηαι, . . .*[11]

(*g*). In the letter itself the following passage, which has been already quoted (p. 111), is found :

'The Delians sent a deputation to the geometers who were staying with Plato at Academia, and requested them to solve the problem [of the duplication of the cube] for them. While they were devoting themselves without stint of labour to the work, and trying to find two mean propor-

planets, and in particular their retrograde and stationary appearances, and also to one of the *spirics* of Perseus, each of which curves has the form of the lemniscate. The passage in Xenophon is as follows :—'Ἱππασίαν δ' ἐπαινοῦμεν τὴν πέδην καλουμένην· ἐπ' ἀμφοτέρας γὰρ τὰς γνάθους στρέφεσθαι ἐθίζει. Καὶ τὸ μεταβάλλεσθαι δὲ τὴν ἱππασίαν ἀγαθὸν, ἵνα ἀμφότεραι αἱ γνάθοι καθ' ἑκάτερον τῆς ἱππασίας ἰσάζωνται. Ἐπαινοῦμεν δὲ καὶ τὴν ἑτερομήκη πέδην μᾶλλον τῆς κυκλοτεροῦς. *Ibid.*, cap. 3 : Τούς γε μὴν ἑτερογνάθους μηνύει μὲν καὶ ἡ πέδη καλουμένη ἱππασία, . . . This curve was named πέδη from its resemblance to the form of the loop of the wire in a snare, which was in fact that of a figure of 8. Some writers have given a different, and, to me it seems, not a correct, interpretation of the origin of this term. Mr. Gow, for example, (*A Short History of Greek Mathematics*, p. 184) says : 'Lastly, Eudoxus is reported to have invented a curve which he called ἱπποπέδη, or "horse fetter," and which resembled those hobbles which Xenophon describes as used in the riding school.' In the next page Mr. Gow says : 'Eudoxus somehow used this curve in his description of planetary motions, . . .' This is not correct : the two curves were of a similar form—that of the lemniscate—and, therefore, the same name was given to each ; but they differed widely geometrically, and were quite distinct from each other. See Knoche and Maerker, *op. cit.* p. 14, *sq.* ; and Schiaparelli, *le Sfere Omocentriche*, &c., p. 32, *sq.*

10 Proclus, ed. Friedlein, pp. 111, 112.

11 Archimedes, ex. rec. Torelli, p. 146 ; Archim., *Opera*, ed. Heiberg., vol. III., p. 112.

tionals between the two given lines, Archytas of Tarentum is said to have discovered them by means of his semi-cylinders, and Eudoxus by means of the so called *curved lines.* It was the lot of all these men to be able to solve the problem with satisfactory demonstration, while it was impossible to apply their methods practically so that they should come into use; except, to some small extent and with difficulty, that of Menaechmus.'[12]

(*h*). The solution of the *Delian Problem* by Menaechmus is also noticed by Proclus in his *Commentary on the Timaeus of Plato* :—'How then, two straight lines being given, it is possible to determine two mean proportionals, as a conclusion to this discussion, I, having found the solution of Archytas, will transcribe it, choosing it rather than that of Menaechmus, because he makes use of the conic lines, and also rather than that of Eratosthenes, because he employs the application of a scale.'[13]

(*i*). The solutions of Menaechmus—of which there are two—have been handed down by Eutocius in his Commentary on the Second Book of the Treatise of Archimedes *On the Sphere and Cylinder*, and will be given at length below.[14]

(*j*). We learn from Plutarch that 'Plato blamed Eudoxus, Archytas, and Menaechmus, and their School, for endeavouring to reduce the duplication of the cube to instrumental and mechanical contrivances; for in this way [he said] the whole good of geometry is destroyed and perverted, since it backslides into the things of sense, and does not soar and try to grasp eternal and incorporeal

[12] Archim., ex. rec. Torelli, p. 144; *ibid.*, ed. Heiberg, vol. III., pp. 104, 106.

[13] Proclus *in Platonis Timæum*, p. 149 in libro III. (ed. Joann. Valder, Basel, 1534). I have taken this quotation and reference from Max. C. P. Schmidt, *die fragmente des Mathematikers Menaechmus*, Philologus, XLII., p. 75. Heiberg, (Archim. *Opera*, vol. III., Praefatio v.) also gives this passage, but his reference is to p. 353, ed. Schneider.

[14] Archim., ed. Torelli, p. 141, *sq.*; Archim. *Opera*, ed. Heiberg, vol. III., p. 92, *sq.*

images; through the contemplation of which God is ever God.'[15]

The same thing is repeated by Plutarch in his *Life of Marcellus* as far as Eudoxus and Archytas are concerned; but in this passage Menaechmus, though not mentioned by name, is, it seems to me, referred to. The passage is:—'The first who gave an impulse to the study of mechanics, a branch of knowledge so prepossessing and celebrated, were Eudoxus and Archytas, who embellish geometry by means of an element of easy elegance, and underprop, by actual experiments and the use of instruments, some problems, which are not well supplied with proof by means of abstract reasonings and diagrams; that problem (for example) of two mean proportional lines, which is also an indispensable element in many drawings:—and this they each brought within the range of mechanical contrivances, by applying certain instruments for finding mean proportionals (μεσογράφους) taken from curved lines and sections (καμπύλων γραμμῶν καὶ τμημάτων). But, when Plato inveighed against them with great indignation and persistence as destroying and perverting all the good there is in geometry, which thus absconds from incorporeal and intellectual to sensible things, and besides employs again such bodies as require much vulgar handicraft: in this way *mechanics* was dissimilated and expelled from geometry, and being for a long time looked down upon by philosophy, became one of the arts of war.'[16]

15 Plut., *Quaest. Conviv.*, lib. VIII., *Q.* ii. 1; *Opera*, ed. Didot, vol. IV., p. 876.

16 Ibid., *Vita Marcelli*, c. 14, sec. 5; Plut., *Opera*, ed. Didot, vol. I., pp. 364, 5. The words καμπύλων γραμμῶν in this passage refer to the curves of Eudoxus (see pp. 132 and 140); τμημάτων refers to the solution of Archytas, and also, in my judgment, to the conic sections. Instead of τμημάτων we should, no doubt, expect to meet τομῶν; but Plutarch was not a mathematician, and the word, moreover, occurs in a biographical work: to this may be added, that in one of the *Definitions* of Heron (*Def.* 91, p. 26, ed. Hultsch) we find τμῆμα used for section.

(*k*). Theon of Smyrna relates that 'he [Plato] blames those *philosophers* who, identifying the stars, as if they were inanimate, with spheres and their circles, introduce a multiplicity of spheres, as Aristotle thinks fit to do, and amongst the *mathematicians*, Menaechmus and Callippus, who introduced the system of deferent and restituent spheres (οἱ τὰς μὲν φερούσας, τὰς δὲ ἀνελιττούσας εἰσηγήσαντο).'[17]

The solutions of Menaechmus referred to in (*i*) are as follows :—

'As Menaechmus.

'Let the two given straight lines be α, ε; it is required to find two mean proportionals between them.—

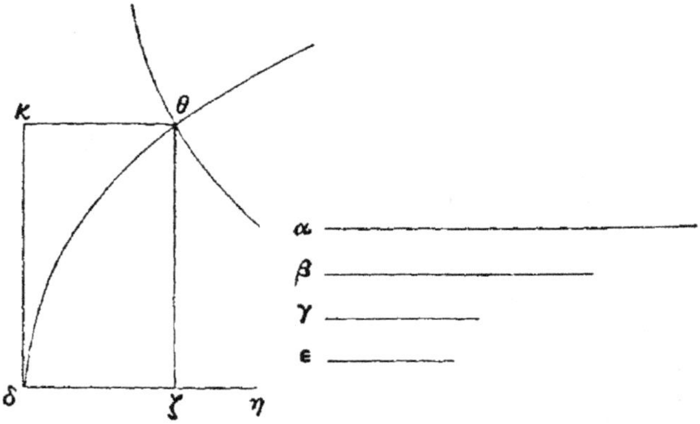

'Let it be done, and let them be β, γ: and let the straight line δη, given in position and limited in δ, be laid

[17] Theonis Smyrnaei Platonici, *Liber de Astronomia*, ed. Th. H. Martin, pp. 330, 332, Paris, 1849. The σφαῖραι ἀνελίττουσαι were, according to this hypothesis, spheres of opposite movement, which have the object of neutralising the effect of other enveloping spheres (Aristot., Met. XII., c. viii., ed. Bekker, p. 1074a). This modification of the system of concentric spheres of Eudoxus is attributed to Aristotle, but we infer from this passage of Theon of Smyrna that it was introduced by Menaechmus (Theon. Smyrn., *Liber de Astron.*, Dissertatio, p. 59). Simplicius, however, in his *Commentary* on Aristotle, *de Caelo* (*Schol.* in Aristot., Brandis, p. 498b), ascribes this modification to Eudoxus himself. Martin (*l. c.*) thinks it probable that this hypothesis was put forward by Menaechmus in his work on Plato's *Republic*, with reference to the description of the distaff of the Fates in the tenth book.

down; and at δ let $\delta\zeta$, equal to the straight line γ, be placed on it, and let the line $\theta\zeta$ be drawn at right angles, and let $\zeta\theta$, equal to the line β, be laid down: since, then, the three straight lines a, β, γ are proportional, the rectangle under the lines a, γ, is equal to the square on β: therefore the rectangle under the given line a and the line γ, that is the line $\delta\zeta$, is equal to the square on the line β, that is to the square on the line $\zeta\theta$; therefore the point θ lies on a parabola described through δ. Let the parallel straight lines $\theta\kappa$, $\delta\kappa$ be drawn: since the rectangle under β, γ is given (for it is equal to the rectangle under a, ϵ), the rectangle $\kappa\theta\zeta$ is also given: the point θ, therefore, lies on a hyperbola described with the straight lines $\kappa\delta$, $\delta\zeta$ as asymptotes. The point θ is therefore given; so also is the point ζ.

'The synthesis will be as follows:—

'Let the given straight lines be a, ϵ, and let the line $\delta\eta$ be given in position and terminated at δ; through δ let a parabola be described whose axis is $\delta\eta$ and parameter a. And let the squares of the ordinates drawn at right angles to $\delta\eta$ be equal to the rectangles applied to a, and having for breadths the lines cut off by them to the point δ. Let it [the parabola] be described, and let it be $\delta\theta$, and let the line $\delta\kappa$ [be drawn and let it] be a perpendicular; and with the straight lines $\kappa\delta$, $\delta\zeta$ as asymptotes, let the hyperbola be described, so that the lines drawn from it parallel to the lines $\kappa\delta$, $\delta\zeta$ shall form an area equal to the rectangle under a, ϵ: it [the hyperbola] will cut the parabola: let them cut in θ, and let perpendiculars $\theta\kappa$, $\theta\zeta$, be drawn. Since, then, the square on $\zeta\theta$ is equal to the rectangle under a and $\delta\zeta$, there will be: as the line a is to $\zeta\theta$, so is the line $\zeta\theta$ to $\zeta\delta$. Again, since the rectangle under a, ϵ is equal to the rectangle $\theta\zeta\delta$, there will be: as the line a is to the line $\zeta\theta$, so is the line $\zeta\delta$ to the line ϵ: but the line a is to the line $\zeta\theta$, as the line $\zeta\theta$ is to $\zeta\delta$. And, therefore:

as the line a is to the line $\zeta\theta$, so is the line $\zeta\theta$ to $\zeta\delta$, and the line $\zeta\delta$ to ϵ. Let the line β be taken equal to the line $\theta\zeta$, and the line γ equal to the line $\delta\zeta$; there will be, therefore: as the line a is to the line β, so is the line β to the line γ, and the line γ to ϵ: the lines a, β, γ, ϵ are, therefore, in continued proportion; which was required to be found.

OTHERWISE.

'Let $a\beta$, $\beta\gamma$ be the two given straight lines [placed] at right angles to each other; and let their mean proportionals be $\delta\beta$, $\beta\epsilon$, so that, as the line $\gamma\beta$ is to $\beta\delta$, so is the

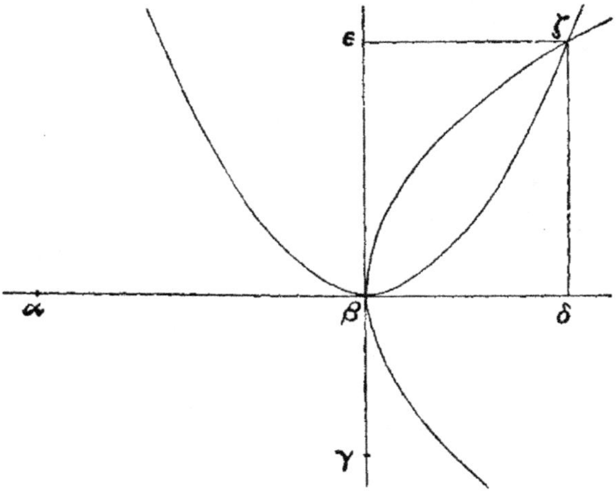

line $\beta\delta$ to $\beta\epsilon$, and the line $\beta\epsilon$ to βa, and let the perpendiculars $\delta\zeta$, $\epsilon\zeta$ be drawn. Since then there is: as the line $\gamma\beta$ is to $\beta\delta$, so is the line $\beta\delta$ to $\beta\epsilon$; therefore the rectangle $\gamma\beta\epsilon$, that is, the rectangle under the given straight line [$\gamma\beta$] and the line $\beta\epsilon$ will be equal to the square on $\beta\delta$, that is [the square] on $\epsilon\zeta$: since then the rectangle under a given line and the line $\beta\epsilon$ is equal to the square on $\epsilon\zeta$, therefore the point ζ lies on a parabola described about the axis $\beta\epsilon$. Again, since there is: as the line $a\beta$ is to $\beta\epsilon$, so is the line $\beta\epsilon$ to $\beta\delta$, therefore the rectangle $a\beta\delta$, that is, the rectangle under the given straight line [$a\beta$] and the line $\beta\delta$, is equal to the square on $\epsilon\beta$, that is [the square] on $\delta\zeta$; the point ζ, therefore, lies on a parabola described about the axis $\beta\delta$: but it [the point ζ] lies also on another given [parabola] described about [the axis] $\beta\epsilon$: the point ζ is therefore given; as are also the perpendiculars $\zeta\delta$, $\zeta\epsilon$: the points δ, ϵ are, therefore, given.

The synthesis will be as follows:—

'Let $\alpha\beta$, $\beta\gamma$ be the two given lines placed at right angles to each other, and let them be produced indefinitely from the point β: and let there be described about the axis $\beta\epsilon$ a parabola, so that the square on any ordinate [$\zeta\epsilon$] shall be equal to the rectangle applied to the line $\beta\gamma$ with the line $\beta\epsilon$ as height. Again, let a parabola be described about $\delta\beta$ as axis, so that the squares on its ordinates shall be equal to rectangles applied to the line $\alpha\beta$. These parabolas cut each other: let them cut at the point ζ, and from ζ let the perpendiculars $\zeta\delta$, $\zeta\epsilon$ be drawn. Since then, in the parabola, the line $\zeta\epsilon$, that is, the line $\delta\beta$, has been drawn, there will be: the rectangle under $\gamma\beta$, $\beta\epsilon$ equals the square on $\beta\delta$: there is, therefore: as the line $\gamma\beta$ is to $\beta\delta$, so is the line $\delta\beta$ to $\beta\epsilon$. Again, since in the parabola the line $\zeta\delta$, that is, the line $\epsilon\beta$, has been drawn, there will be: the rectangle under $\delta\beta$, $\beta\alpha$ equals the square on $\epsilon\beta$: there is, therefore: as the line $\delta\beta$ is to $\beta\epsilon$, so is the line $\beta\epsilon$ to $\beta\alpha$; but there was: as the line $\delta\beta$ is to $\beta\epsilon$, so is the line $\gamma\beta$ to $\beta\delta$: and thus there will be, therefore: as the line $\gamma\beta$ is to $\beta\delta$, so is the line $\beta\delta$ to $\beta\epsilon$, and the line $\beta\epsilon$ to $\beta\alpha$; which was required to be found.'

Eutocius adds—'The parabola is described by means of a compass (διαβήτου) invented by Isidore of Miletus, the engineer, our master, and described by him in his Commentary on the Treatise of Heron *On Arches* (καμαρικῶν).'

We have, therefore, the highest authority—that of Eratosthenes, confirmed by Geminus, (*e*) and (*f*)—for the fact that Menaechmus was the discoverer of the three conic sections, and that he conceived them as sections of the cone. We see, further, that he employed two of them, the parabola and the rectangular hyperbola, in his solutions of the Delian problem. We learn, however, from a passage of Geminus, quoted by Eutocius in his Commentary on the *Conics* of Apollonius, which has already been referred to in another connection (p. 11) that these names, *parabola* and *hyperbola*, are of later origin, and were given to these curves by Apollonius:—

'But what Geminus says is true, that the ancients (*οἱ παλαιοί*), defining a cone as the revolution of a right-angled triangle, one of the sides about the right angle remaining fixed, naturally supposed also that all cones were right, and that there was one section only in each—in the right-angled one, the section now called a *parabola*, in the obtuse-angled, the *hyperbola*, and in the acute-angled the *ellipse*; and you will find the sections so named by them. As then the original investigators (*ἀρχαίων*) observed the two right angles in each individual kind of triangle, first in the equilateral, again in the isosceles, and lastly in the scalene; those that came after them proved the general theorem as follows:—"The three angles of every triangle are equal to two right angles." So also in the sections of a cone; for they viewed the so called "section of the right-angled cone" in the right-angled cone only, cut by a plane at right angles to one side of the cone; but the section of the obtuse-angled cone they used to show as existing in the obtuse-angled cone; and the section of the acute-angled cone in the acute-angled cone; in like manner in all the cones drawing the planes at right angles to one side of the cone; which also even the original names themselves of the lines indicate. But, afterwards, Apollonius of Perga observed something which is universally true—that in every cone, as well right as scalene, all these sections exist according to the different application of the plane to the cone. His contemporaries, admiring him on account of the wonderful excellence of the theorems of conics proved by him, called Apollonius the "*Great Geometer.*" Geminus says this in the sixth book of his *Review of Mathematics.*'[18]

The statement in the preceding passage as to the original names of the conic sections is also made by Pappus, who says, further, that these names were given to them by Aristaeus, and were subsequently changed by Apollonius to those which have been in use ever since.[19] In the writings of Archimedes, moreover, the conic sections are always called by their old names, and thus this statement of Geminus is indirectly confirmed.[20]

It is much to be regretted that the two solutions of Menaechmus have not been transmitted to us in their original form. That they have been altered, either by Eutocius or by some author whom he followed, appears not only from the employment in these solutions of the terms parabola and hyperbola, as has been already frequently pointed out,[21] but much more from the fact that the lan-

[18] Apollonii *Conica*, ed. Halleius, p. 9.

[19] Pappi Alexand. *Collect.* VII., ed. Hultsch, p. 672, *sq.* Mr. Gow (*op. cit.*), p. 186, note, says: 'That Menaechmus used the name "section of right-angled cone," etc., is attested by Pappus, VII. (ed. Hultsch), p. 672.' This is not correct; the name of Menaechmus does not occur in Pappus.

[20] Heiberg (*Nogle Puncter af de graeske Mathematikeres Terminologi*, Kjobenhavn, 1879, p. 3.) points out that 'only in three passages is the word ἔλλειψις found in the works of Archimedes, but everywhere it ought to be removed as a later interpolation, as Nizze has already asserted.' These passages are: 1°. περὶ κωνοειδέων, ed. Torelli, p. 270, ed. Heiberg, vol. I. pp. 324, 326; 2°. *ibid.* Tor. p. 272, Heib., *id.* p. 332, l. 22; 3°. *ibid.* Tor. p. 273, Heib. *id.* p. 334, l. 5. Heiberg, moreover, calls attention to a passage where Eutocius (*Comm.* to Archimedes, περὶ σφαίρας καὶ κυλίνδρου, II. ed. Tor. p. 163, ed. Heib., vol. III., p. 154, l. 9) attributes to Archimedes a fragment he had discovered, containing the solution of a problem which requires the application of conic sections, among other reasons because in it their original names are used.

[21] First, as far as I know, by Reimer, *Historia problematis de cubi duplicatione*, Gottingae, 1798, p. 64, note.

guage used in them is, in its character, altogether that of Apollonius.[22]

Let us now examine whether any inference can be drawn from the previous notices as to the way in which Menaechmus was led to the discovery of his curves. This question has been considered by Bretschneider,[23] whose hypothesis as to the course of the inquiry is very simple and quite in accordance with what we know of the state of geometry at that time.

We have seen that the right cone only was considered, and was conceived to be cut by a plane perpendicular to a side; it is evident, moreover, that this plane is at right angles to the plane passing through that side and the axis of the cone. We have seen, further, that if the vertical angle of the cone is right, the section is the curve, of which the fundamental property—expressed now by the equation $y^2 = px$—was known to Menaechmus. This being premised, Bretschneider proceeds to show how this property of the parabola may be obtained in the manner indicated.

Let DEF be a plane drawn at right angles to the side AC of the right cone whose vertex is A, and circular base BFC; and let the triangle BAC (right-angled at A) be the section of the cone made by the plane drawn through AC and the axis of the cone. Let the plane DEF cut the cone in the curve DKF, and the plane BAC in the line DE. If, now, through any point J of the line DE a plane HKG be drawn parallel to the base BFC of the cone, the section of the cone made by this plane will be a circle, whose plane will be at right angles to the plane BAC; to which plane the plane of the section DKF is also perpendicular; the

[22] *e.g.* παραβολή, ὑπερβολή, ἀσύμπτωτος, ἄξων, ὀρθία πλευρά. The original name for the asymptotes, αἱ ἔγγιστα, is met with in Archimedes, *de Conoidibus*, &c. (αἱ ἔγγιστα τᾶς τοῦ ἀμβλυγωνίου κώνου τομᾶς, ed. Heiberg, vol. I., p. 276, l. 22; and again, αἱ ἔγγιστα εὐθεῖαι, κ. τ. λ., *id.*, p. 278, l. 1). See Heiberg, *Nogle Punct.*, &c., p. 11.

[23] Bretsch., *Geom. vor Eukl.*, p. 156, *sq.*

line JK of intersection of these two planes will then be at right angles to the plane BAC, and, therefore, to each of the lines HG and DE in that plane. Let now the line DL

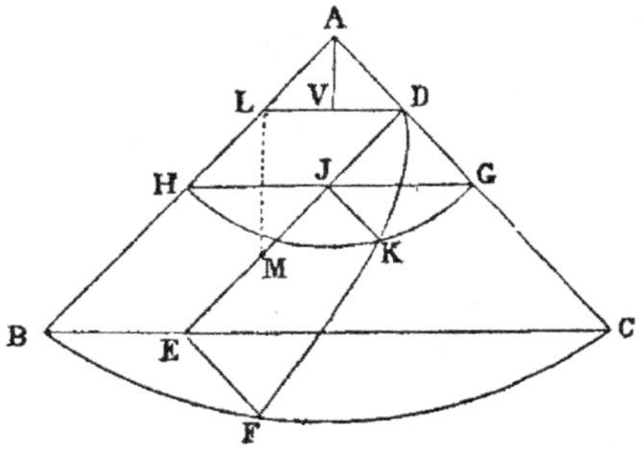

be drawn parallel to HG, and the line LM at right angles to LD. In the semicircle HKG the square on JK is equal to the rectangle HJG, that is, to the rectangle under LD and JG, or, on account of the similar triangles JDG and DLM, to the rectangle under DJ and DM. The section of the right-angled cone, therefore, is such that the square on the ordinate KJ is equal to the rectangle under a given line DM and the abscissa DJ.

Bretschneider proceeds then to the consideration of the sections of the acute-angled and obtuse-angled cones, and investigates the manner in which Menaechmus may have been led to the discovery of properties similar to those which he had known in the semicircle, and found in the case of the section of the right-angled cone.

Let a plane be drawn perpendicular to the side AC of an acute-angled cone, and let it cut the cone in the curve DKE, and let the plane through AC and the axis cut the cone in the triangle BAC. Through any point J of the line DE let a plane be drawn parallel to the base of the cone, cutting the cone in the circle HKG, whose plane will be at right angles to the plane BAC, to which plane the plane of the section DKE is also perpendicular. The line JK of intersection of these two planes will then be at right angles to the plane BAC; and, therefore, to each of the lines HG and DE in that plane, draw LD and EF parallel to HG, and at the point L draw a perpen-

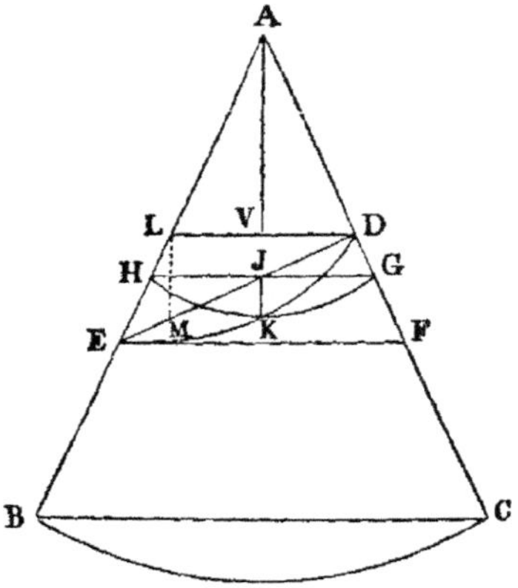

dicular to LD, intersecting DE in the point M. We have then

$$HJ : JE :: LD : DE,$$

$$JG : JD :: EF : DE;$$

therefore,

$$HJ \,.\, JG : JE \,.\, JD :: LD \,.\, EF : DE^2.$$

But, on account of the similar triangles DEF and DLM,

$$EF : DE :: MD : LD.$$

Hence we get

$$HJ \,.\, JG : JE \,.\, JD :: MD : DE.$$

But in the semicircle HKG

$$JK^2 = HJ \,.\, JG;$$

therefore,

$$JK^2 : JE \,.\, JD :: MD : DE,$$

that is, the square of the ordinate JK is to the rectangle under EJ and JD in a constant ratio.

The investigation in the case of the section of the obtuse-angled cone is similar to the above.

Bretschneider observes that the construction given for MD in the preceding investigations is so closely connected with the position of the plane of section DKE at right angles to the side AC, that it could scarcely have escaped the observation of Menaechmus.

This hypothesis of Bretschneider, as to the properties of the conic sections first perceived by Menaechmus, which properties he employed to distinguish his curves from each other, seems to me to be quite in accordance as well with the state of geometry at that time as with the place which Menaechmus occupied in its development.

A comparison of these investigations with the solution of Archytas (see p. 111, *sq.*) will show, as there stated, that 'the same conceptions are made use of, and the same course of reasoning is pursued' in each (p. 115):

In each investigation two planes are perpendicular to an underlying plane; and the intersection of the two planes is a common ordinate to two curves lying one in each plane. In one of the intersecting planes the curve is in each case a semicircle, and the common ordinate is, therefore, a mean proportional between the segments of its diameter. So far the investigation is the same for all. Now, from the consideration of the figure in the underlying plane—which is different in each case—it follows that:—in the first case—the solution of Archytas—the ordinate in the second intersecting plane is a mean proportional between the segments of its base, whence it is inferred that the extremity of the ordinate in this plane also lies on a semicircle; in the second case—the section of the right-angled cone—the ordinate is a mean proportional between a given straight line and the abscissa; and, lastly, in the third case—the section of an acute-angled cone—the ordinate is proportional to the geometric mean between the segments of the base.

So far, it seems to me, we can safely go, but not farther. From the first solution of Menaechmus, however, it has been generally inferred that he must have discovered the asymptotes of the hyperbola, and have known the property of the curve with relation to these lines, which property we now express by the equation $xy = a^2$. Menaechmus may have discovered the asymptotes; but, in my judgment, we are not justified in making this assertion, on account of the fact, which is undoubted, that the solutions of Menaechmus have not come down to us in his own words. To this may be added that the words *hyperbola* and *asymptotes* could not have been used by him, as these terms were unknown to Archimedes.

From the passage in the letter of Eratosthenes at the end of extract (*g*), coupled with the statement of Plutarch (*j*), Bretschneider infers that it is not improbable that Menaechmus invented some instrument for drawing his curves.[24] Cantor considers this interpretation as not impossible, and points out that there is in it no real contradiction to the observation in Eutocius concerning the description of the parabola by Isidore of Miletus.[25] Bretschneider adds that if Menaechmus had found out such an instrument it could never have been in general use, since not the slightest further mention of it has come down to us. It appears to me, however, that it is more probable that Menaechmus constructed the parabola and hyperbola by points, though this supposition is rejected by Bretschneider on the ground that such a construction would be very tedious. On the other hand, it seems to me that the words of Eratosthenes would apply very well to such a procedure. We know, on the authority of Eudemus (see p. 24), that 'the inventions concerning the application of areas'—on which, moreover, the construction by points of the curves $y^2 = px$ and $xy = a^2$ depend—'are ancient, ἀρχαῖα, and are due to the Pythagoreans':[26] it may be fairly inferred, then, that problems

[24] Bretsch., *Geom. vor Eukl.*, p. 162.
[25] *Gesch. der. Math.*, p. 211.
[26] Proclus, ed. Friedlein, p. 419.

of application were frequently solved by the Greeks. And we have the very direct testimony of Proclus in the passage referred to, that the inventors of these constructions applied them also to the arithmetical solution of the corresponding problems. It is not surprising, therefore, to find—as M. Paul Tannery[27] has remarked—Diophantus constantly using the expression *παραβάλλειν παρὰ* in the sense dividing.[28]

The extracts from Proclus (*b*), (*c*), and (*d*) are interesting as showing that Menaechmus was not only a discoverer in geometry, but that questions on the philosophy of mathematics also engaged his attention.

In the passages (*c*) and (*d*), moreover, the expression *οἱ περὶ Μέναιχμον μαθηματικοί* occurs—precisely the same expression as that used by Iamblichus with reference to

[27] *De la Solution Géométrique des Problèmes du Second Degré avant Euclide* (Mémoires de la Société des Sciences phys. et nat. de Bordeaux, tom. IV., 2ᵉ Série, 1882, p. 409. Tannery (Bulletin des Sciences math. et astron., tom. IV., 1880, p. 309) says that we must believe that Menaechmus made use of the properties of the conic sections, which are now expressed by the equation between the ordinate and the abscissa measured from the vertex, for the construction of these curves by points.

[28] In a paper published in the Philologus, *Griechische und römische mathematik* (Phil. XLIII., 1884, pp. 474–5), Heiberg puts forward views which differ widely from those stated above. He holds:—that it is not certain that Menaechmus contrived an apparatus for the delineation of the conic sections: that the only meaning which can be attached to Plato's blame (*j*) is, that Archytas, Eudoxus, and Menaechmus had employed, for the duplication of the cube, curves which could not be constructed with the rule and compass; and that the passage of Eratosthenes merely says that the curves of Menaechmus could be constructed, and not that he had found an apparatus for the purpose. Heiberg says, moreover, that it cannot be doubted that the Pythagoreans solved, by means of the application of areas, the equations, which we now call the vertical equations of the conic sections: but while admitting this, he holds that there is no ground for inferring thence that these equations were employed for the description of the conic sections by points; and says that such a description by points runs counter to the whole spirit of Greek geometry. On the other hand, it seems to me that Tannery is right in believing that the *quadratrix* of Deinostratus (the brother of Menaechmus), or of Hippias, the contemporary of Socrates, was constructed in this manner (see Bulletin des Sciences math. et astron., *pour l'histoire des lignes and surfaces courbes dans l'Antiquité*, tom. VII., p. 279). Moreover, the construction of the parabola and rectangular hyperbola by points depends on the simplest problems of application of areas—the *παραβολή* without the addition of the *ὑπερβολή* or *ἔλλειψις*.

Eudoxus (p. 134)—and we observe that in (*d*) this expression stands in contrast with *οἱ περὶ Σπεύσιππον*, which is met with in the same sentence. From this it follows that Menaechmus had a school, and that it was looked on as a *mathematical* rather than as a *philosophical* school. Further, we have seen that Theon of Smyrna makes a similar distinction between Aristotle on the one side and Menaechmus and Callippus on the other (*k*). Lastly, we learn from Simplicius that Callippus of Cyzicus, a pupil of Polemarchus—the friend of Eudoxus—went subsequently to Athens and lived with Aristotle, with whom he conferred in order to revise and complete the inventions of Eudoxus.[29]

When these statements are put together, and taken in conjunction with the fact mentioned by Ptolemy, that Callippus made astronomical and meteorological observations at the Hellespont,[30] we are, I think, justified in assuming that the reference in each is to the School of Cyzicus, founded by Eudoxus, whose successors were—Helicon (probably), Menaechmus, Polemarchus, and Callippus.

From the passages of Plutarch referred to in (*j*) we see that Plato blamed Archytas, Eudoxus, and Menaechmus for reducing the duplication of the cube to mechanical contrivances. On the other hand the solution of this problem,

[29] The passage is in the *Commentary* of Simplicius on the Second Book of Aristotle *de Caelo*, and is as follows:—εἴρηται καὶ ὅτι πρῶτος Εὔδοξος ὁ Κνίδιος ἐπέβαλε ταῖς διὰ τῶν ἀνελιττουσῶν καλουμένων σφαιρῶν ὑποθέσεσι, Κάλλιππος δὲ ὁ Κυζικηνὸς Πολεμάρχῳ συσχολάσας τῷ Εὐδόξου γνωρίμῳ, καὶ μετ' ἐκεῖνον εἰς Ἀθήνας ἐλθών, τῷ Ἀριστοτέλει συγκατεβίω, τὰ ὑπὸ τοῦ Εὐδόξου εὑρεθέντα σὺν τῷ Ἀριστοτέλει διορθούμενός τε καὶ προσαναπληρῶν.—*Scholia* in Aristot. Brandis, p. 498b. Callippus and Polemarchus, as Boeckh has remarked, could not have been fellow-pupils of Eudoxus: Callippus, who flourished *circ.* 330 B.C., was too young. The meaning of the passage must be as stated above. Boeckh conjectures that Polemarchus was about twenty years older than Callippus. See *Sonnenkreise*, p. 155.

[30] φάσεις ἀπλανῶν ἀστέρων καὶ συναγωγὴ ἐπισημασιῶν, Ptolemy, ed. Halma, Paris, 1819, p. 53.

attributed to Plato, and handed down by Eutocius, is purely mechanical. Hence grave doubts have arisen as to whether this solution is really due to Plato. These doubts are increased if reference be made to the following authorities:—

Eratosthenes, in his letter in which the history of the Delian problem is given, refers to the solutions of Archytas, Eudoxus, and Menaechmus, but takes no notice of any solution by Plato, though mentioning him by name; Theon of Smyrna also, quoting a writing of Eratosthenes entitled 'The Platonic,' relates that the Delians sent to Plato to consult him on this problem, and that he replied that the god gave this oracle to the Delians, not that he wanted his altar doubled, but that he meant to blame the Hellenes for their neglect of mathematics and their contempt of geometry.[31] Plutarch, too, gives a similar account of the matter, and adds that Plato referred the Delians, who implored his aid, to Eudoxus of Cnidus, and Helicon of Cyzicus, for its solution.[32] Lastly, Jo. Philoponus, in his account of the matter, agrees in the main with Plutarch, but in Plato's answer to the Delians he omits all reference to others.[33]

Cantor, who has collected these authorities, sums up the evidence, and says the choice lies between—1° the assumption that Plato, when blaming Archytas, Eudoxus, and Menaechmus, added, that it was not difficult to execute the doubling of the cube mechanically; that it could be effected by a simple machine, but that this was not geometry; or 2° the rejection, as far as Plato is concerned, of the communication of Eutocius, on the ground of the statements of Plutarch and the silence of Eratosthenes; or lastly, 3° the admission that a contradiction exists here which we have not sufficient means to clear up.[34]

[31] Theon. Smyrn. *Arithm.*, ed. de Gelder, Lugdun. Bat. 1827, p. 5.

[32] Plutarch, *de Genio Socratis*, *Opera*, ed. Didot., vol. III., p. 699.

[33] Jo. Philop. *ad Aristot. Analyt. post.*, I. vii.

[34] Cantor, *Gesch. der Math.*, p. 202.

The fact that Eratosthenes takes no notice of the solution of Plato seems to me in itself to be a strong presumption against its genuineness. When, however, this silence is taken in connection with the statements of Plutarch, that Plato referred the Delians to others for the solution of their difficulty, and also that Plato blamed the solutions of the three great geometers, who were his contemporaries, as mechanical—a condemnation quite in accordance, moreover, with the whole spirit of the Platonic philosophy—we are forced, I think, to the conclusion that the sources from which Eutocius took his account of this solution are not trustworthy. This inference is strengthened by the fact, that the source, from which the solution given by Eudoxus of the same problem was known to Eutocius, was so corrupt that it was unintelligible to him, and, therefore, not handed down by him.[35]

The solution atttributed to Plato is as follows:—

'As Plato.

'Two straight lines being given to find two mean proportionals in continued proportion.

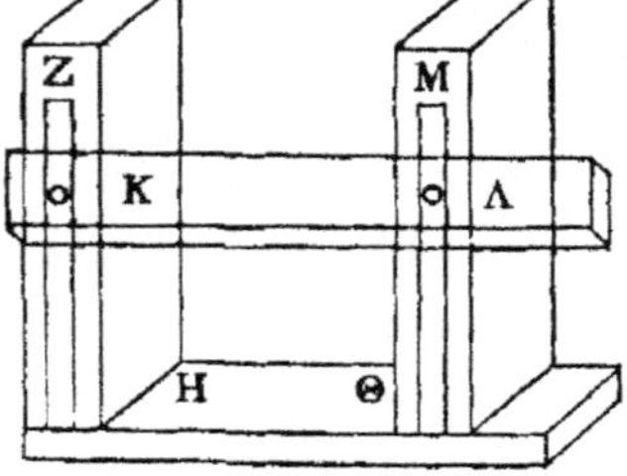

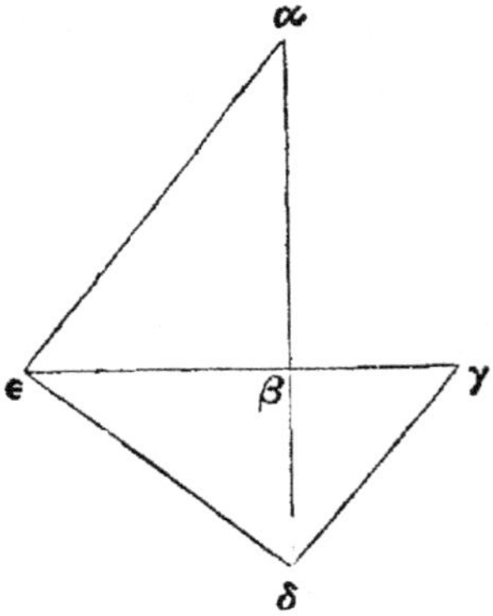

'Let the two given straight lines $\alpha\beta$, $\beta\gamma$, between which it is required to find two mean proportionals, be at right

[35] See p. 140.

angles to each other. Let them be produced to δ, ϵ. Now let there be constructed a right angle ZHΘ, and in either leg, as ZH, let a ruler KΛ be moved in a groove which is in ZH, so as to remain parallel to HΘ. This will take place if we imagine another ruler connected with ΘH and parallel to ZH, as ΘM. For the upper surfaces of the rulers ZH, ΘM being furrowed with grooves shaped like a dove-tail, in these grooves tenons connected with the ruler KΛ being inserted, the motion of the ruler KΛ will be always parallel to HΘ. This being arranged, let either leg of the angle, as HΘ, be placed in contact with the point γ, and let the angle and the ruler be moved so far that the point H may fall on the line $\beta\delta$, whilst the leg HΘ is in contact with the point γ, and the ruler KΛ be in contact with the line $\beta\epsilon$ at the point K, but on the other side with the point α: so that, as in the diagram, a right angle be placed as the angle $\gamma\delta\epsilon$, but the ruler KΛ have the position of the line $\epsilon\alpha$. This being so, what was required will be done; for the angles at δ and ϵ being right, there will be the line $\gamma\beta$ to $\beta\delta$, as the line $\delta\beta$ to $\beta\epsilon$, and the line $\epsilon\beta$ to $\beta\alpha$.'[36]

The instrument is in fact a gnomon, or carpenter's square, with a ruler movable on one leg and at right angles to it, after the manner of a shoemaker's size-stick.

If this solution be compared with the second solution of Menaechmus it will be seen that the arrangement of the two given lines and their mean proportionals is precisely the same in each, and that, moreover, the analysis must also be the same. Further, a reference to the solution of Archytas (see pp. 111 and 114 (*b*)) will show that the only geometrical theorems made use of in the solution attributed to Plato were known to Archytas. Hence it seems

[36] Archim., ed. Torelli, p. 135; Archim. *Opera*, ed. Heiberg, vol. III., p. 66, *sq*. I have taken the diagrams used in this solution and that of Menaechmus from Heiberg's edition of Archimedes.

to me that it may be fairly inferred that this solution was subsequent to that of Menaechmus, as his solution was to that of Archytas. This, so far as it goes, is in favour of the first supposition of Cantor given above.

On account of the importance of the subject treated of here, I will state briefly my views on the matter in question:—Menaechmus was led by the study of the solution of Archytas, in the manner given above, to the discovery of the curve whose property (σύμπτωμα) is that now defined by the equation $y^2 = px$. Starting from this, he arrived at the properties of the sections of the acute-angled and of the obtuse-angled right cones, which are analogous to the well-known property of the semicircle—the ordinate is a mean proportional between the segments of the diameter. Having found the curve defined by the property, that its ordinate is a mean proportional between a given line and the abscissa, Menaechmus saw that by means of two such curves the problem of finding two mean proportionals could be solved, as given in the second of his two solutions, which, I think, was the one first arrived at by him. The question was then raised—Of what practical use is your solution? or, in other words, how can your curve be described?

Now we have seen that, side by side with the development of abstract geometry by the Greeks, the practical art of geometrical drawing, which they derived originally from the Egyptians, continued to be in use: that the Pythagoreans especially were adepts in it, and that, in particular, they were occupied with problems concerning the application (παραβολή) of areas, including the working of numerical examples of the same. Now any number of points, as near to each other as we please, on the curve $y^2 = px$, can be obtained with the greatest facility by this method; and in this manner, I think, Menaechmus traced the curve known subsequently by the name parabola—a name transferred from the *operation* (which was the proper signification of παραβολή) to the result of the operation. We have seen that the same name, παραβολή, was transferred and applied to division, which was also a transference of a name of an operation to its result.

Having solved the problem by the intersection of two parabolas, I think it probable that Menaechmus showed that the practical solution of the question could be simplified by using, instead of one of them, the curve $xy = a^2$, the construction of which by points is even easier than that of the parabola. There is no evidence, however, for the inference that Menaechmus knew that this curve was the same as the one he had obtained as a section of the obtuse-angled cone; or that he knew of the existence of the asymptotes of the hyperbola, and its equation in relation to them.

Let us examine now whether anything can be derived from the sources, which would enable us to fix the time of the Delian deputation to Plato—be it real or fictitious.

We have seen that Sotion, after mentioning that Eudoxus took up his abode at Cyzicus and taught there and in the neighbouring cities of the Propontis, relates that subsequently he returned to Athens accompanied by a great many pupils (*πάνυ πολλοὺς περὶ ἑαυτὸν ἔχοντα μαθητάς*), for the sake, as some say, of annoying Plato, because formerly he had not held him worthy of attention (p. 129). We learn, further, from Apollodorus, that Eudoxus flourished about the hundred and third Olympiad—B.C. 367—and it is probable, as Boeckh thinks, that this time falls in with his residence at Cyzicus. Now the narrative of Plutarch—that Plato referred the Delians to Eudoxus and Helicon for the solution of their difficulty—points to the time of the visit of Eudoxus and his pupils to Athens, for—1° as we know from Sotion, Plato and Eudoxus had not been on good terms; and 2° it is not probable that, before this visit, Helicon, who was a native of Cyzicus and a pupil of Eudoxus, as we learn from the spurious 13th *Epistle of Plato*, had become famous or was known to Plato. Boeckh assumes, no doubt rightly, that the visit of Eudoxus and his pupils to Athens, and their sojourn there, took place a few years later than Ol. 103, 1—B.C. 367; so that it occurred between the second and third visits of Plato to Sicily (368 B.C. and 361 B.C.).[37] To this time, therefore, he refers the remarkable living and working together at the Academy of eminent men, who were distinguished in mathematics and astronomy, according to the report of Eudemus as handed down by Proclus. Now, amongst those named there we find Eudoxus himself, his pupil Menaechmus, Deinostratus—the brother of Menaechmus —and Athenaeus of Cyzicus;[38] to these must be added Helicon of Cyzicus—more distinguished as an astronomer than a mathematician—who was recommended to Dionysius by Plato,[39] and who was at the court of Dionysius in company with Plato at the time of his third visit to Syracuse.[40]

I quite agree with Boeckh in thinking that all the pupils of Eudoxus and the citizens of Cyzicus, whom we find at Athens at that time—even though they are not expressly named as pupils of Eudoxus—belonged to the school of Cyzicus; and I have no doubt that to these illustrious Cyzicenians the fame of the Academy—so far at least as mathematics and astronomy are concerned —is chiefly due.[41] It is noteworthy that Aristotle, at the

[37] Boeckh, *Sonnenkreise*, pp. 156, 157.

[38] See Introduction, pp. 4, 5.

[39] *Epist.* Plat., xiii.

[40] Plutarch, *Dion.*

[41] Zeller says: 'Among the disciples of Plato who are known to us, we find many more foreigners than Athenians: the greater number belong to that eastern portion of the Greek world which since the Persian war had fallen chiefly under the influence of Athens. In the western regions, so far as these were at all ripe for philosophy, Pythagoreanism, then in its first and most

time of this visit, so famous and so important in consequence of the impetus thereby given to the mathematical sciences, had recently joined the Academy, and was then a young man; and it is easy to conceive the profound impression made by Eudoxus and his pupils on a nature like that of Aristotle; and an explanation is thus afforded as well of the great respect which he entertained for Eudoxus, as of the cordial relations which existed later between him and the mathematicians and astronomers of the school of Cyzicus.[42]

flourishing period, most probably hindered the spread of Platonism, despite the close relation between the two systems' (*Plato and the Older Academy*, E. T, p. 553, *sq.*). Zeller gives in a note a list of Plato's pupils, in which all the distinguished men of the School of Cyzicus are placed to the credit of the Academy.

[42] Aristotle was born in the year 384 B.C., and went to Athens 367 B.C. Aafter the death of Plato (B.C. 347), Aristotleleft Athens and went to Atarneus in Mysia, where his friend Hermias was *dynast*. When he was there he may have renewed his relations with the distinguished men of the School of Cyzicus, which was not far distant. It is quite possible that Menaechmus may have been recommended as mathematical teacher to Alexander the Great by Aristotle; and we have seen that Callippus of Cyzicus, who had been a pupil of Polemarchus, went to Athens to hold a conference with Aristotle on the system of Eudoxus, with the view of revising and completing it.

CHAPTER VII.

THE SUCCESSORS OF EUDOXUS.

II. DEINOSTRATUS.*

Deinostratus.—The Quadratrix, its generation and characteristic property.—Use of this curve to solve the two famous problems: 1°. the Trisection of an Angle; 2°. the Quadrature of the Circle.—Enunciation and proof of the property of the Quadratrix on which this second employment of the Curve depends.—Criticisms of Sporus and Pappus on the Curve and on this use of it.—Theorems required for, and axioms used in, the proof of this property of the Curve.—These axioms are substantially the same as the well-known principles assumed by Archimedes.—The Problem solved by means of the Quadratrix is the rectification of the Quadrant, and is a complement to the work of Eudoxus.—Is the Hippias mentioned by Proclus in connection with the Quadratrix the same as the Sophist Hippias of Elis?

DEINOSTRATUS was brother of Menaechmus, and is mentioned by Eudemus, together with Amyclas and Menaechmus, as having made the whole of geometry more perfect (p. 4).

The only notice of his work which has come down to us is contained in the following passage of Pappus:—

'For the quadrature of the circle a certain curve[1] was employed by Deinostratus, Nicomedes, and some other more recent geometers, which has received its name from

* The two works announced in the note (p. 150) have appeared: Autolyci *de Sphaera quae movetur Liber, de ortibus et occasibus Libri duo:* una cum scholiis antiquis e libris manuscriptis edidit Latina interpretatione et commentariis instruxit Fridericus Hultsch, Lipsiae, 1885; *Diophantos of Alexandria: a Study in the History of Greek Algebra*, by T. L. Heath, Cambridge, 1885.

The following works have also been published: Euclidis *Elementa*, edidit et Latine interpretatus est J. L. Heiberg, Dr. Phil., vol. IV. libros XI.-XIII. continens, Lipsiae, 1885; *die Lehre von den Kegelschnitten im Altertum* von Dr. H. G. Zeuthen, erster halbband, Kopenhagen, 1886.

[1] γραμμή. The Greeks had no special name for 'a curve.'

the property that belongs to it; for it is called by them the quadratrix (τετραγωνίζουσα), and its generation is as follows:

'Let a square αβγδ be assumed, and about the centre γ let the quadrant[2] βεδ be described, and let the line γβ be moved so that the point γ remain fixed, and the point β be borne along the quadrant βεδ: again, let the straight line βα, always remaining parallel to the line γδ, accompany the point β while it is borne along the line βγ; and let the line γβ, moving uniformly, pass over the angle βγδ—that is, the point β describe the quadrant βεδ—in the same time in which the straight line βα traverses the line βγ—that is, the point β is borne along βγ. It will evidently

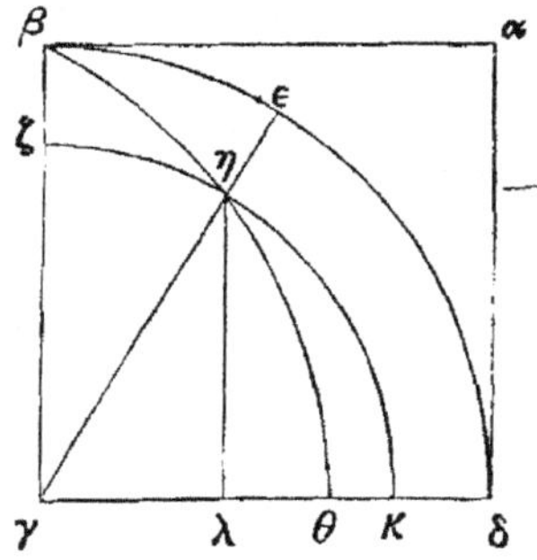

happen that each of the lines γβ and βα will coincide simultaneously with the straight line γδ. Such then being the motion, the straight lines βα, βγ in their motion will cut one another in some point, which always changes its place with them; by which point, in the space between the straight lines βγ, γδ, and the quadrant βεδ, a certain curve concave towards the same side such as βηθ, is described; which indeed seems to be useful for finding a square, which shall be equal to a given circle. But its characteristic property (ἀρχικὸν αὐτῆς σύμπτωμα) is this:—if

[2] περιφέρεια, arc. 'Ex recentiorum usu Graecam περιφέρειαν, id est partem aliquam totius circuli circumferentiae, Ernestum Nizze, Theodosii interpretem, secuti plerumque *arcum* interpretati sumus.' (Autolycus, *op. cit.*, Praefatio, pp. xiv, xv.)

any line, as $\gamma\eta\epsilon$, be drawn to the circumference, as the whole quadrant $\beta\epsilon\delta$ is to the arc $\epsilon\delta$, so is the straight line $\beta\gamma$ to $\eta\lambda$; for this is evident from the generation of the curve.'[3]

Pappus, has, moreover, transmitted to us the property of the quadratrix, from which it received its name, together with the proof. It is as follows :—

'If $\alpha\beta\gamma\delta$ be a square, and $\beta\epsilon\delta$ be the quadrant about the centre γ, and the line $\beta\eta\theta$ be the quadratrix described as in the manner given above; it is proved that: as the quadrant $\delta\epsilon\beta$ is to the straight line $\beta\gamma$, so is $\beta\gamma$ to the straight line $\gamma\theta$. For if it is not, the quadrant $\delta\epsilon\beta$ will be to the line $\beta\gamma$ as $\beta\gamma$ to a line greater than $\gamma\theta$, or to a lesser.

'In the first place let it be, if possible [as $\beta\gamma$], to a greater line $\gamma\kappa$; and about the centre γ let the quadrant $\zeta\eta\kappa$ be described, cutting the curve at the point η; let the perpendicular $\eta\lambda$ be drawn, and let the joining line $\gamma\eta$ be produced to the point ϵ. Since then: as the quadrant $\delta\epsilon\beta$ is to the straight line $\beta\gamma$, so is $\beta\gamma$—that is $\gamma\delta$—to the line $\gamma\kappa$, and as $\gamma\delta$ is to $\gamma\kappa$, so is the quadrant $\beta\epsilon\delta$ to the quadrant $\zeta\eta\kappa$ (for the circumferences of circles are to each other as their diameters),[4] it is evident that the quadrant $\zeta\eta\kappa$ is equal to the straight line $\beta\gamma$. And since, on account of the property of the curve, there is: as the quadrant $\beta\epsilon\delta$ is to the arc $\epsilon\delta$, so is $\beta\gamma$ to $\eta\lambda$; and therefore: as the quadrant $\zeta\eta\kappa$ is to the arc $\eta\kappa$, so is the straight line $\beta\gamma$ to the line $\eta\lambda$. And it has been shown that the quadrant $\zeta\eta\kappa$ is equal to the straight line $\beta\gamma$; therefore the arc $\eta\kappa$ will be equal to the straight line $\eta\lambda$, which is absurd. Therefore it is not true that: as the quadrant $\beta\epsilon\delta$ is to the straight line $\beta\gamma$, so is $\beta\gamma$ to a line greater than $\gamma\theta$.'

'Further, I say, that neither is it to a line less than $\gamma\theta$. For, if possible, let it be to $\gamma\kappa$, and about the centre γ let

[3] Pappi *Collect.*, ed. Hultsch, vol. I., pp. 250, 252.

[4] 'Hoc theorema extat V. propos. II et VIII. propos. 22; simul autem scriptor tacite efficit circulorum arcus quibus aequales anguli insistunt inter se esse ut radios.' (*Ibid.*, p. 257, *n.*)

the quadrant $\zeta\mu\kappa$ be described, and let the line $\kappa\eta$ be drawn at right angles to the line $\gamma\delta$, cutting the quadratrix at the point η, and let the joining line $\gamma\eta$ be produced to the point ϵ. In like manner then to what has been proved above, we show that the quadrant $\zeta\mu\kappa$ is equal to the straight line $\beta\gamma$, and that: as the quadrant $\beta\epsilon\delta$ is to the arc $\epsilon\delta$—that is, as the quadrant $\zeta\mu\kappa$ to the arc $\mu\kappa$—so is the straight line $\beta\gamma$ to the line $\eta\kappa$. From which it is evident that the arc $\mu\kappa$ is equal to the straight line $\kappa\eta$, which is absurd. Therefore it is not true that: as the quadrant $\beta\epsilon\delta$ is to the straight line $\beta\gamma$, so is $\beta\gamma$ to a line less than $\gamma\theta$. Neither is it to a greater, as has been proved above: therefore it is to the line $\gamma\theta$ itself.'[5]

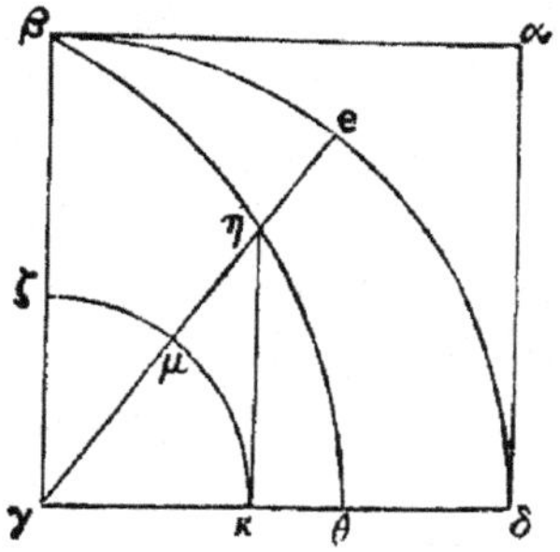

Pappus continues—'This also is evident, that if a third proportional be taken to the straight lines $\theta\gamma$, $\gamma\beta$, the straight line [thus found] will be equal to the quadrant $\beta\epsilon\delta$; and four times this line will be equal to the circumference of the whole circle. But the straight line, which is equal to the circumference of a circle, being found, it is evident that a square equal to the circle itself can be easily constructed: for the rectangle under the perimeter of a circle and its radius is double of the circle, as Archimedes proved.'[6]

[5] Pappi *Collect.*, vol. I., pp. 256, 258.

[6] 'Paulo aliis verbis Pappus id theorema enuntiat atque ipse Archimedes circuli dimens. propos. I: πᾶς κύκλος ἴσος ἐστὶ τριγώνῳ ὀρθογωνίῳ, οὗ ἡ μὲν ἐκ τοῦ κέντρου ἴση μιᾷ τῶν περὶ τὴν ὀρθήν, ἡ δὲ περίμετρος τῇ λοιπῇ. (*Ibid.*, p. 259, *n.* 2.)

Pappus also relates that Sporus justly found fault with this curve, for two reasons :—

1. 'It takes for granted the very thing for which the quadratrix is employed ; for it is not possible to make one point move from β to γ along the straight line $\beta\gamma$ in the same time that another point moves along the quadrant $\beta\epsilon\delta$, unless the ratio of the straight line to the quadrant is first known, inasmuch as it is necessary that the rates of the motions should be to each other in the same ratio.

2. 'The extremity of the curve which is employed for the quadrature of the circle—that is, the point in which the quadratrix cuts the straight line $\gamma\delta$—is not found ; for when the straight lines $\gamma\beta$, $\beta\alpha$, being moved, are brought simultaneously to the end of their motion, they coincide with the line $\gamma\delta$, and no longer cut one another—for the cutting ceases before the coincidence with the line $\alpha\delta$, which intersection on the other hand is taken as the extremity of the curve, in which it meets the straight line $\alpha\delta$: unless, perhaps, some one might say that the curve should be considered as produced—just as we suppose that straight lines are produced—as far as $\alpha\delta$; but this by no means follows from the principles laid down : but in order that this point θ may be assumed, the ratio of the quadrant to the straight line must be presupposed.'

He then adds, that 'unless this ratio is given, one should not—trusting to the authority of the inventors—accept a curve, which is rather of a mechanical kind (τὴν γραμμὴν μηχανικωτέραν πως οὖσαν).[7]

Sporus was a mathematician whose solution of the Delian problem has been handed down by Eutocius in his Commentary on the treatise of Archimedes *On the Sphere and Cylinder* ;[8] this solution, he tells us, is the same as that of Pappus, which precedes it in Eutocius, and which is also

[7] Pappi *Collect.*, vol. I., pp. 252, 254.
[8] Archim., *Opera*, ed. Heiberg, vol. III., pp. 90, 92.

given by Pappus himself in the third and eighth books of his *Collections*.[9] M. Paul Tannery thinks that Sporus was the teacher, or an elder fellow-pupil of Pappus, and places him towards the end of the third century of our era; and, further, he identifies him with Porus (Sporus) of Nicaea, the author of a collection entitled Ἀριστοτελικὰ Κηρία (see p. 59), which contained, according to M. Tannery, extracts from mathematical works relating to the quadrature of the circle and the duplication of the cube, as also a compilation in relation to the *Meteorologics* of Aristotle. M. Tannery is of opinion, moreover, that the historical works of Eudemus were driven out of the field at an early period by compilations from them; that the *History of Geometry*, in particular, did not survive the fourth century; and that this collection of Sporus was the principal source from which Pappus, Simplicius, and Eutocius derived their information concerning these two famous geometrical problems.[10]

In any case, it seems to me probable that a valuable fragment of the *History of Geometry* of Eudemus is preserved in the extracts from Pappus given above, whether they have been taken by Pappus from that *History*, or derived second-hand through Sporus [Porus].

On examining the demonstration of the property of the quadratrix given above, we see that the following theorems are required for it:—

(*a*). The circumferences of circles are to each other as their diameters.

(*b*). The arcs of two concentric circles, which subtend the same angle at their common centre, are to each other as the quadrants of those circles.

[9] Pappi *Collect.*, vol. I., p. 64, *sq.*; vol. III. p. 1070, *sq.*

[10] *Sur les fragments d' Eudème de Rhodes relatifs à l'histoire des mathématiques*; also, *sur Sporos de Nicée*; Annales de la Faculté des Lettres de Bordeaux, pp. 70–76, 257–261, 1882. Cf. *Pour l'histoire des lignes et surfaces courbes dans l'antiquité*. Bulletin des Sciences mathém. et astronom., 2e série, tom. VII.

This theorem is an immediate consequence of Euclid, VI. 33:—

(*c*). In equal circles, angles at the centre have the same ratio to each other as the arcs on which they stand.

We see, further, that the following assumptions are made in the proof:—

1°. An arc of a circle less than a quadrant is greater than the perpendicular let fall from one of its extremities on the radius drawn through the other;

2°. And is less than the tangent drawn at one extremity of the arc to meet the radius produced through the other.

We notice, moreover, that the proof is indirect; and it is, indeed, as Cantor has remarked, the first of the kind with which we meet.[11] We have seen, however, that Eudoxus must have been familiar with this method of reasoning (p. 139); and we know that Autolycus of Pitane, in Aeolis, who was a contemporary of Deinostratus, makes use of the argument *ad absurdum* (ὅπερ ἐστὶν ἄτοπον, or ἀδύνατον), in many propositions of his book περὶ κινουμένης σφαίρας.[12]

We see, too, that the investigation of Deinostratus, which gives a graphical solution of the determination of the ratio of the circumference of a circle to its diameter, is a complement to the work of Eudoxus, for the problem which was solved by means of the quadratrix arose naturally from the theorem that *circles are to each other as the squares on their diameters.*

It is to be observed, then, in the first place, that the problem which is solved above by means of the quadratrix is, in reality, the rectification of the quadrant, and that it is taken for granted that the quadrature of the circle—from which the name of the curve is derived—follows from

[11] Cantor, *Gesch. der Math.*, p. 213. [This remark is not correct, see p. 43, and *n*. 64.]

[12] Autolycus, *op. cit.*, pp. 12, 4; 14, 7; 24. 14; 32, 4; 8, 17; 22, 1.

its rectification. Secondly, we see that in order to make this inference the theorem—the area of a circle is equal to one-half the rectangle under the circumference, or four times the quadrant, and the radius—must be assumed. This theorem is equivalent to the first proposition of Archimedes, *Dimensio circuli*, referred to above. Lastly, it is noteworthy that the rectification of the quadrant is obtained by means of principles which are substantially the same as those assumed by Archimedes, and adopted by all geometers, ancient and modern.[13]

It seems to be a legitimate inference from this that these axioms must be referred back to Deinostratus, and most probably to Eudoxus.

Pappus, no doubt, in two places—v., prop. 11, and VIII., prop. 22—proves that the circumferences of circles are to each other as their diameters,[14] and, in each place, makes the proof depend on the theorem cited above. He adds, however, in the former proposition :—'The same may be proved without assuming that the rectangle under the diameter of a circle and its periphery is four times the circle. For the similar polygons, which are inscribed in circles, or circumscribed about them, have perimeters which have the same ratio to each other as the radii of the circles, so that also the circumferences of circles are to each other as their diameters.'

Bretschneider thinks that the criticisms of Sporus are not of much importance, and says that they only come to this :—'That the quadratrix cannot be constructed geo-

[13] 'Nous partirons, pour la solution de ce problème [de la rectification des courbes], du principe d' *Archimède*, adopté par tous les géomètres anciens et modernes, suivant lequel deux lignes courbes, ou composées de droites, ayant leurs concavités tournées du même côté et les mêmes extrémités, celle qui renferme l'autre est la plus longue. D'où il suit qu'un arc de courbe tout concave du même côté, est plus grand que sa corde, et en même temps moindre que la somme des deux tangentes menées aux deux extrémités de l'arc, et comprises entre ces extrémités et leur point d'intersection.'—Lagrange, *Théorie des Fonctions Analytiques*, p. 218. Paris, 1813.

[14] Pappi *Collect.*, vol. I., pp. 334, 336; vol. III., pp. 1104, 1106.

metrically, but is obtained only mechanically by means of a series of points, which must then be joined by a steady stroke of the free hand.[15] It seems to me, however, that these criticisms are just; and that Sporus and Pappus are right in maintaining that the description of the curve assumes the very thing for which the quadratrix is employed.[16]

Bretschneider shows that the theorem from which the quadratrix derives its name can be easily obtained by the infinitesimal method, 'by means of the proportion $\beta\epsilon\delta : \gamma\delta :: \epsilon\delta : \eta\lambda$, from the observation that the nearer the radius $\gamma\epsilon$ approaches to $\gamma\delta$, the more nearly does the sector $\gamma\epsilon\delta$ approach to a triangle similar to the triangle $\gamma\lambda\eta$; and therefore, for the limiting case, where $\gamma\epsilon$ and $\gamma\delta$ coincide, the ratio $\epsilon\delta : \eta\lambda$ actually passes over into that of $\gamma\delta : \gamma\theta$.' He adds :—'Such considerations have often served the old geometers as means for their discoveries, but are never used as proofs. The latter are always given through the *reductio ad absurdum*, which, indeed, allows no trace of the way followed in the inquiry to be recognised.'[17] This observation is both just and important.

The same remark has been made by M. P. Laffitte, who points out that, in the establishment of any truth, there are two parts (or operations) which, he says, have not been hitherto sufficiently distinguished:

1°. The invention or the discovery of the proposition.
2°. Its proof.

And he further observes, that, after the discovery has been

[15] Bretsch., *Geom. vor Eukl.*, p. 96.

[16] 'Various other modes might be found of making either of these curves [the quadratrix of Deinostratus and the quadratrix of Tschirnhausen] square the circle; but the fact is that the description of the curves themselves assumes the point which their use is to determine.'—*English Cyclopædia*, sub. v., *Quadratrix*.

[17] Bretsch., *Geom. vor Eukl.*, p. 154.

arrived at, the proof is often furnished by the method *ex absurdo.*[18]

In the third chapter (p. 93, *sq.*) I gave reasons in support of Hankel's opinion that the Hippias referred to by Proclus, in connection with the quadratrix, is not Hippias of Elis. As I mentioned, however, in giving them, I had not then read Cantor's defence of the common opinion; but, on reading it subsequently, I was much struck with the force of his arguments, and introduced them in a note—the only course then open to me. M. Paul Tannery, in a Paper, the first part of which was published in the *Bulletin des Sciences mathématiques et astronomiques*, Octobre, 1883, and entitled, 'Pour l'histoire des lignes et surfaces courbes dans l'antiquité,'[19] has criticised the reasons advanced by me against the common opinion:—

With reference to argument 1°, he replies:—'This omission is sufficiently explained by the discredit under which the sophists laboured in the eyes of Eudemus; and the list in question presents a much more remarkable one—that of Democritus.'

With reference to 2°, he says:—'This observation is not accurate. An indefinite number of points of the quadratrix, as near as one wishes, may be obtained by the ruler and compass; and it is doubtful whether the ancients sought any other process for the construction of this curve.' M. Tannery continues:—'The authority of Diogenes Laertius is, moreover, so much the less acceptable, inasmuch as he speaks in express terms of the solution of the Delian problem by Archytas. Now, Eutocius (*Archimedes*, ed. Torelli, pp. 143–144) has preserved to us, on the one hand, this solution, in which there is not any employment of an instrument; and, on the other hand (p. 145), a letter, in

18 P. Laffitte, *les Grands Types de l'Humanité*, vol. II., p. 308, *sq.*; p. 328, *sq.*
19 *Bulletin des Sc. math. et astron.*, 2e série, tom. VII. 1, p. 279, *sq.*

which Eratosthenes states that, "if Archytas, Eudoxus, &c., were able to prove the accuracy of their solutions, they could not realise them manually and practically, except, to a certain extent, Menaechmus, but in a very troublesome way."' (Cf. p. 111.)

'The *Mesolabe* of Eratosthenes is, in fact, the oldest instrument of which the employment for a geometrical construction is known. This text indicates that, before Menaechmus, people were not engrossed with the practical tracing of curves; whilst the inventor of the conic sections would have tried, more or less, to resolve this question for the lines which he had discovered.'

As to these observations of M. Tannery, I admit that Diogenes Laertius is not a safe guide in mathematics, as indeed I noticed in the first chapter (p. 10, *n.* 12). In quoting him, I certainly did not mean to convey that, in my opinion, Archytas had actually traced the curve, used in his solution of the Delian problem, by any mechanical means; and I agree with M. Tannery that the letter of Eratosthenes is quite decisive on that point. At the same time it is evident that the conception of a curve being traced by means of motion is contained in the solution of Archytas, to whom, along with Philolaus, his master, and Eudoxus, his pupil, the first notions of mechanics are attributed. And with respect to the quadratrix itself, although, as M. Tannery remarks, an indefinite number of points on the quadratrix, as near as one wishes, can be obtained with the ruler and compass, yet the conception of motion is no less involved in the nature and very definition of the curve.

In reply to my observation 3°, M. Tannery says:—'The divergence of the accounts given by Proclus and by Pappus is easily explained by the difference of the sources from which they drew. All that the former says of curves is undoubtedly borrowed from Geminus, an author of the first century before the Christian era; and his language proves that Geminus was acquainted with a writing of Hippias on the quadratrix, and regarded him as the inventor of this curve, though he was aware that Nicomedes also was engaged with it.' M. Tannery continues :—'As to Pappus, he quotes Geminus only *apropos* of the works of Archimedes on Mechanics. He does not appear to have borrowed anything from him for geometry, particularly in the part which is concerned with curved lines and surfaces;' and adds :—'One can scarcely doubt but that Sporus was the source from which Pappus has derived what he says on the quadratrix.' We have noticed this above.

With reference to 4°, M. Tannery says:—'The existence of the Hippias referred to in it is by no means proved, for the writing in question seems to be only a pure fancy; but in any case it is impossible to think of any geometer posterior to Geminus, or even, as it seems to me, to Nicomedes.'

The suggestion which I made concerning Hippias, the contemporary of Lucian, was thrown out by me without sufficient consideration in reply to the observation of Montucla. Later, I became aware of the ideal character of that writing, and that it was the work of a *pseudo*-Lucian.[20]

The result of the whole discussion seems to be: that the quadratrix was invented, probably by Hippias of Elis, with the object of trisecting an angle, and was originally employed for that purpose; that subsequently Deinostratus used the curve for the quadrature of the circle, and that its name was thence derived. This seems to be Cantor's view of the matter.[21] M. Tannery tells us that he, too, had at first interpreted the passage of Pappus in the same way as Cantor; but that, on further consideration, he thinks that

[20] See Zeller, *History of Greek Philosophy from the earliest period to the time of Socrates*, vol. II., p. 422, *n.* 2, E.T.

[21] Cantor, *Gesch. der Math.*, pp. 167 and 212.

it is open to grave objections. He says:—'In the first place, the text of Geminus in Proclus clearly supposes that the name of the curve had been given to it by its inventor, Hippias. On the other hand, it is evident that the practical use of the curve implies the construction of a model cut in a square, having the quadratrix in place of the hypotenuse, and which could be applied, like our *protractor*, to the figures under consideration. Consequently, the determination of the intersection of the curve with the axis at once becomes necessary; and the problem is not, in reality, so difficult that we should think that Hippias was incapable of perceiving its relation to the quadrature of the circle. Finally, the fame of this last problem was at the time sufficiently great to lead Hippias to borrow from it the name of his curve, rather than from the problem which he had, without any doubt, considered in the first place.'[22]

These views of M. Tannery seem to me to be inadmissible, and are indeed quite inconsistent with what we know of Greek geometry (see *supra*, p. 95, *sq.*; p. 138, *sq.*).[23] The problem solved by means of the quadratrix must, as stated above, be regarded as the natural complement of the work of Eudoxus; and it is significant, therefore, that the solution was effected by Deinostratus, who probably was his pupil. Nor does the finding of the point of intersection of the curve with the axis necessarily involve the determination of π; for, as seems to be suggested by Pappus, the required point might be regarded as determined by the production of the curve. Should it be said that the theorem required for the determination of π was obtained first by the infinitesimal method, I would reply that it was

[22] *Bull. des Sc. math. et astron.*, 2e série, VII., I., p. 281.

[23] Cf. Heiberg, *Griechische und römische Mathematik*, Philologus, 1884, *Jahresberichte*, p. 474: 'Während Hankel, p. 121, ff. die exhaustionsmethode auf Hippokrates zurückgehen liess, und Cantor, p. 209, die möglichkeit zugibt, hebt Allman, Greek Geometry &c., II., p. 221 ff. [p. 95, sq. *supra*] mit recht hervor, dass wir nicht berechtigt sind, diese methode für älter als Eudoxus zu halten.'

not likely that this was done by Hippias of Elis, who was a senior contemporary of Democritus. If, then, the text in Proclus supposes that the name of the curve had been given to it by its inventor, it follows, in my opinion, that this could not have been Hippias of Elis. I am, however, on the whole, disposed to accept Cantor's view as given above.

CHAPTER VIII.

THE SUCCESSORS OF EUXODUS.

III. ARISTAEUS.

Aristaeus.—He was probably a senior contemporary of Euclid.—Passages in Pappus relating to him quoted.—What is ὁ ἀναλυόμενος τόπος? It was treated of by Aristaeus, Euclid, and Apollonius of Perga.—List of the books which are contained in it.—Aristaeus wrote five books on 'Solid Loci'.—Names given by him to the three Conic Sections.—These names were changed by Apollonius to those still in use.—Reason for the change and significance of the new names.—Aristaeus wrote a book on the 'Comparison of the five Regular Solids'.—He was one of the most important Geometers before Euclid.—Discussion as to whether Aristaeus wrote one work only on the Conic Sections, namely, the 'Solid Loci' in five books, or whether he wrote also the 'Elements of Conics' likewise in five books.—Theorem of Aristaeus.—Enumeration of the Theorems required in its proof as given by Hypsicles.—Simple proof of the 'Theorem of Aristaeus'.—Retrospect.—Relation of the work done by Aristaeus to that of Archytas, and his successors.

PAPPUS has preserved the name, and given some account of the work, of one other great geometer, who was a predecessor, and probably a senior contemporary of Euclid—ARISTAEUS the elder. We have no details whatever of his life.

The passages in Pappus relating to him are as follows:—

(*a*) 'That which is called ὁ ἀναλυόμενος [τόπος],[1] that is,

[1] [τόπος] ὁ καλούμενος ἀναλυόμενος. τόπος, '*locus*, i.e. quicquid aliqua mathematicarum parte comprehenditur: ὁ ἀστρονομούμενος τόπος, VI. 474, 3; ὁ ἀναλυόμενος τόπος, VII. 672, 4.' *Index Graecitatis*, Pappi *Collect.*, voluminis III. tomus II., p. 114. 'ὁ ἀναλυόμενος τόπος, *locus de resolutione*, id est *doctrina analytica*.' *Ibid.* sub voce, ἀναλύειν, p. 5. Compare what Marinus says on the same subject in his Commentary on the *Data* of Euclid:

'What is the value of the treatise about *Data*?'

'The *datum* having been divided in a general way, and as far as is sufficient the department of mathematics which treats of analysis, is, in short, a certain peculiar matter prepared for those who, having gone through the elements, wish to acquire the power of solving problems proposed to them in the construction of lines ; and it is useful for this purpose only. It has been treated of by three men—Euclid, the author of the Elements, Apollonius of Perga, and Aristaeus the elder—and proceeds by the method of analysis and synthesis.'[2]

Pappus, having defined analysis and synthesis, proceeds to give a complete list of the books, arranged in order, which are contained in the τόπος ἀναλυόμενος. He enumerates thirty-three books in all, amongst which we find 'five books of Aristaeus on *solid loci*' (Ἀρισταίου τόπων στερεῶν πέντε): the remaining books, with the exception of two by Eratosthenes concerning *means* (περὶ μεσοτήτων δύο), were written by Euclid and Apollonius.[3]

(*b*) '[These plane problems, then, are found in the τόπος ἀναλυόμενος, and are set out first, with the exception of the *means* of Eratosthenes ; for these come last. Next to plane problems order requires the consideration of solid problems. Now, they call solid problems, not only those which are proposed in solid figures, but also those which, not being capable of solution by plane loci, are solved by means of the three conic lines, and so it is necessary to write first concerning these. Five books of the *Elements of Conics* were first published by the elder Aristaeus, which

for the present need, the next point is to state the utility of treatment of the subject. This also is one of those things which have their result in relation to something else. For the knowledge of this is necessary in the highest degree for τὸν ἀναλυόμενον τόπον as it is called ; and how much value ὁ ἀναλυόμενος τόπος has in mathematical science, and the kindred science of optics and music, has been defined elsewhere, and that analysis is the discovery of a proof, and that it helps us to the discovery of things similar, and that it is more important to possess the analytical faculty than to have many proofs of particular things.' *Euclidis Data*, ed. Cl. Hardy, p. 13. Cf. Pappi *Collect.*, Appendix, p. 1275.

[2] Pappi *Collect.*, VII., vol. II., p. 634.

[3] *Ibid.*, p. 636.

were written in a compendious manner, inasmuch as those who took up the study of them were now able to follow him.][4]

(*c*) 'Apollonius, completing Euclid's four books of conics, and adding four others, published eight volumes of conics. But Aristaeus, who wrote the five volumes of *solid loci*, which have come down to the present time, in continuation of the *conics* ('Αρισταῖος δέ, ὃς γέγραφε τὰ μέχρι τοῦ νῦν ἀναδιδόμενα στερεῶν τόπων τεύχη έ συνεχῆ τοῖς κωνικοῖς), called [as also did those before Apollonius] the first of the three conic lines, the section of the acute-angled cone; the second, the section of the right-angled cone; the third, the section of the obtuse-angled cone. But since in each of these three cones, according to the way in which it is cut, these three lines exist, Apollonius, as it appears, felt a difficulty as to why at all his predecessors distinguished by name the section of an acute-angled cone, which might also be that of the right-angled and obtuse-angled cone; and, again, the section of the right-angled cone, which might also be that of the acute-angled and the obtuse-angled cone. Wherefore, changing the names, he called that which had been named the section of the acute-angled cone, the ellipse; the section of the right-angled cone, the parabola; and the section of the obtuse-angled cone, the hyperbola—each from a certain peculiar property. For the rectangle applied to a certain straight line in the section of the acute-angled cone is deficient (ἐλλείπει) by a square; in the section of the obtuse-angled cone it is excessive (ὑπερβάλλει) by a square; finally, in the section of the right-angled cone the rectangle applied (παραβαλλόμενον) is neither deficient nor excessive.

'[But this happened to Aristaeus, since he did not perceive that, according to a peculiar position of the plane cutting the cone, the three curves exist in each of the cones,

[4] Pappi *Collect.*, p. 672. 'τὰ μέν—γεγραμμένα, interpolatori tribuit Hultsch.' The spaced words are supplied in translation.

which curves he named from the peculiarity of the cone. For if the cutting plane be drawn parallel to one side of the cone, one only of the three curves is generated, and that one always the same, which Aristaeus named the section of that so cut cone.]'[5]

(*d*) 'But as to what he [Apollonius] says in the third book, that the locus with three or four lines has not been completed by Euclid—for neither he himself, nor anyone else, could [solve that locus] by those conical [theorems] only which had been proved up to the time of Euclid, as also he himself testifies, saying that it was not possible to complete it without those things which he was compelled to discuss before-hand—[as to this, Euclid, approving of Aristaeus as a worthy mathematician on account of the conics which he had handed down, and not being in haste, nor wishing to lay down anew the same treatment of these subjects (ὁ δὲ Εὐκλείδης ἀποδεχόμενος τὸν Ἀρισταῖον ἄξιον ὄντα ἐφ' οἷς ἤδη παραδεδώκει κωνικοῖς, καὶ μὴ φθάσας ἢ μὴ θελήσας ἐπικαταβάλλεσθαι τούτων τὴν αὐτὴν πραγματείαν)—for he was most kind and friendly to all those who were able to advance mathematics to any extent, as is right, and by no means disposed to cavil, but accurate, and no boaster like this man Apollonius—wrote as much as could be proved by his conics: sc. those of Aristaeus concerning that locus—not attributing any finality to his demonstration, for then it would be necessary to blame him, but, as it is, not at all; since Apollonius also himself, who left many things in his conics unfinished, is not brought to task for it. But he, Apollonius, has been able to add to that locus (τῷ τόπῳ) what was wanting, having been furnished with the ideas by the books already written by Euclid on the same locus (περὶ τοῦ τόπου), and having been for a long time a fellow-pupil of the disciples of Euclid in Alexandria, from which

[5] Pappi *Collect.*, p. 672, l. 18—p. 674, l. 19. 'l. 12. τοῦτο δ' ἔπαθεν (scil. ὁ Ἀρισταῖος)—l. 19. τομὴν interpolatori tribuit Hultsch.' Cf. Proclus, ed. Friedlein, pp. 419, 420. See pp. 164, 165.

source he derived his habit of thought, which is not unscientific. Such is this locus with three or four lines, on which he plumes himself greatly, adding, that he knew that he owed thanks to him who first wrote about it.]'[6]

(*e*) We also learn from Hypsicles that Aristaeus wrote a book on the *Comparison of the five regular solids*, and that it contained the theorem: 'The same circle circumscribes the pentagon of the dodecahedron and the triangle of the icosahedron, these solids being inscribed in the same sphere.' Hypsicles says, further, that 'this theorem is also given by Apollonius in the second edition of his *Comparison of the dodecahedron with the icosahedron*,[7] which is: The surface of the dodecahedron is to the surface of the icosahedron as the dodecahedron itself is to the icosahedron; since the perpendiculars from the centre of the sphere to the pentagon of the dodecahedron and to the triangle of the icosahedron are the same.'[8]

The foregoing extracts lead us to form a high opinion of Aristaeus, and to see that he was one of the most important geometers before Euclid. We have, therefore, great reason to regret the total loss of his writings.

In the passage (*a*) Aristaeus, Euclid, and Apollonius, are named as the three authors on the doctrine of analysis. This passage shows, further, the value that was attached by the ancients to the five books of Aristaeus on *solid loci*, which was one of the works—indeed one of the higher works—included in the τόπος ἀναλυόμενος. From the passage (*b*) it would appear that Aristaeus published also a work on the *elements of conics* in five books—an abridg-

[6] Pappi *Collect.*, p. 676, l. 19—p. 678, l. 15. 'l. 25. ὁ δὲ Εὐκλείδης—p. 678, l. 15, τοιοῦτός ἐστιν, scholiastae cuidam historiae quidem veterum mathematicorum non imperito, sed qui dicendi genere languido et inconcinno usus sit, tribuit Hultsch,' *Ibid.*, p. 677. As Hultsch says, 'the writer of this passage has employed a feeble and awkward manner of expression'; and it is difficult to see the exact meaning of it. The spaced words are supplied in translation.

[7] πέντε σχημάτων σύγκρισις.

[8] Euclid, Book XIV., Prop. 2. This book is in reality the work of Hypsicles.

ment introductory to the study of solid loci. Of his work on *solid loci* it is, moreover, stated in (*c*): Ἀρισταῖος δέ, ὅς γέγραφε τὰ μέχρι τοῦ νῦν ἀναδιδόμενα στερεῶν τόπων τεύχη έ συνεχῆ τοῖς κωνικοῖς. This passage admits of several interpretations:—

1. That the work on *solid loci* was intended as an extension of the theory of conics;

2. Aristaeus first wrote the τόποι στερεοί in five books, and then, to facilitate the study of them, he wrote the κωνικὰ στοιχεῖα—an epitome—also in five books;

3. τοῖς κωνικοῖς might possibly refer to the conics of Euclid.

We learn further from (*c*) that Aristaeus gave to the conic sections their original names, those by which they were known before Apollonius.[9] From (*d*) we learn that Euclid praised the conics of Aristaeus, whom he valued highly, and from the words ἐφ' οἷς ἤδη παραδεδώκει κωνικοῖς, and φθάσας, it has been concluded that he was a predecessor, and probably a senior contemporary of Euclid.[10]

We have seen that the passage (*b*) is regarded by Hultsch as an interpolation. In this Heiberg agrees, and infers thence that Aristaeus wrote only one work on the conic sections—τόποι στερεοί in five books—and holds that the generally received opinion that Aristaeus, besides the five books τόποι στερεοί, had written five more books κωνικὰ στοιχεῖα is not sufficiently well founded. He says: 'The only passage which can be adduced for it, Pappus VII., p. 672, 11: ἣν μὲν οὖν ἀναδεδομένα κωνικῶν στοιχείων πρότερον Ἀρισταίου τοῦ πρεσβυτέρου έ τεύχη, ὡς ἂν ἤδη δυνατοῖς οὖσι τοῖς ταῦτα παραλαμβάνουσιν ἐπιτομώτερον γεγραμμένα, is rightly rejected by Hultsch as not genuine,' and continues, 'It occurs in a perfectly wrong place where Apollonius περὶ νεύσεων is referred to, is objectionable in many respects in

[9] Cf. pp. 164, 165.
[10] J. L. Heiberg, *Studien über Euklid*, p. 85.

point of language, and contains nothing but what a reader of Pappus already would find in him; I believe, therefore, that we, in the words p. 672, 4–14, have a scholium which originally stood in the margin after p. 672, 16, and later fell into the text in a wrong place: the scholiast has then called the five books *τόποι στερεοί*, here incorrectly *στοιχεῖα κωνικά*. And even were the passage genuine (and only misplaced) the probability would be then that Pappus here by *στοιχεῖα κωνικά* had meant the *τόποι*.'[11]

With this conclusion of Heiberg I cannot agree. In the first place, it should be observed that the passages of Pappus enclosed by Hultsch in [ ] are to be considered as interpolations for reasons of style, not of substance. The passage referred to was either written by Pappus himself (as Cantor and others assume), or it originated with an experienced commentator (scholiast), whose statements in other passages also are acknowledged as correct—or, to doubt which there is no occasion; or else these scholia contain remnants of the tradition of the mathematical school of Alexandria, and this tradition must be considered on the whole as correct, so long as the contrary is not proved.[12]

In the next place, Heiberg is not correct in saying that 'it is the only passage which can be adduced for it.' The same statement is made expressly in the text of Pappus himself, a few lines lower down, in the passage quoted above: *'Αρισταῖος δέ, ὃς γέγραφε τὰ μέχρι τοῦ νῦν ἀναδιδόμενα στερεῶν τόπον τεύχη ε̄ συνεχῆ τοῖς κωνικοῖς* (p. 672, l. 20). Heiberg tries to obviate this objection by interpreting

[11] J. L. Heiberg, *l. c.*

[12] It is certain that Pappus had a school. It may, therefore, be assumed that one—or perhaps several—of his pupils had taken notes of his lectures; and that these notes, arising thus from the oral exposition of Pappus himself, were worked out further by his pupils, and formed Commentaries, which were then written on the margin, and subsequently received into the text, of the work which has come down to us as Πάππου συναγωγή. These Commentaries are easily recognised by their style, but as to their contents, they must be considered to be of almost equal authority with the undoubted text of Pappus.

συνεχῆ as meaning: 'which stands in connection with the doctrine of the conic sections—depends on it'. In passage (*d*), moreover, the conics of Aristaeus are, I think, directly referred to in the words: διὰ τῶν ἐκείνου ['Αρισταίου] κωνικῶν. Heiberg, further, says that the interpolation, or scholium, occurs in a perfectly wrong place; but, as he shows, it has to be placed only two lines lower. My view of the matter is that given above, p. 199, 2:—Aristaeus first wrote the *solid loci* in five books, and then, to facilitate the study of them he wrote the *Elements of Conics*—an epitome—also in five books.

The *Conics* of Aristaeus, no doubt, do not appear in the list of books contained in the so-called τόπος ἀναλυόμενος; neither do those of Euclid: they were both replaced by the *Conics* of Apollonius in eight books.

We have seen that Aristaeus wrote a work on the comparison of the five regular solids, and that it contained the theorem: The same circle circumscribes the pentagon of the dodecahedron and the triangle of the icosahedron, these solids being inscribed in the same sphere (*e*).

If we examine the proof of this theorem as given by Hypsicles, we see that it depends on the following theorems:—

1. If a regular pentagon be inscribed in a circle, the square on a side, together with the square on the line subtending two sides of the pentagon, is five times the square on the radius of the circle;

2. If the line subtending two sides of a regular pentagon be cut in extreme and mean ratio, the greater segment is the side of the pentagon. Euclid, XIII. 8;

3. The side of a regular decagon inscribed in a circle is the greater segment of the radius cut in extreme and mean ratio;

4. The square on the side of a regular pentagon inscribed in a circle is equal to the sum of the squares on the

sides of the regular hexagon and decagon inscribed in the same circle. Euclid, XIII. 10;

5. If an equilateral triangle be inscribed in a circle, the square on the side is three times the square on the radius. Euclid, XIII. 12;

6. The square on the diameter of a sphere is three times the square on the side of the inscribed cube. Euclid, XIII. 15;

7. The line subtending two sides of the pentagon of a dodecahedron inscribed in a sphere is the side of the cube inscribed in the same sphere;

This follows from (2) taken with the corollary of XIII. 17: If the side of the cube be cut in extreme and mean ratio, the greater segment is the side of the dodecahedron;

8. The square on the diameter of a sphere is five times the square on the radius of the circle by means of which the iscosahedron is descried—*i. e.* the circle circumscribing the pentagon which forms the base of the five equilateral triangles having for common vertex any vertex of the icosahedron. Euclid, XIII. 16, and Corollary.

From the fact that 'the work of Aristaeus on the *Comparison of the regular solids* is the newest and last that treated, *before* Euclid, of this subject,' Bretschneider infers that 'the contents of the thirteenth book of the Elements is a recapitulation, at least partial, of the work of Aristaeus.'[13] This supposition of Bretschneider receives, I think, great confirmation from the above examination, which shows that the principal propositions in Book XIII. of the Elements are required for the demonstration, as given by Hypsicles, of the theorem of Aristaeus. This theorem, moreover, goes beyond what is contained in the Elements on this subject.

Further, one of the four problems treated of by Pappus in the third book of his *Collection* is the inscription in the sphere of the five regular polyhedra. M. Paul Tannery has

[13] *Geom. vor Eukl.*, p. 171.

thrown out the suggestion that it is probably taken from the *Comparison of the five figures* by Aristaeus the elder, but has given no reasons for his opinion.[14] In support of this conjecture I would put forward that:—

1. Pappus concludes his treatment of the subject by saying that 'from the construction it is evident that the same circle circumscribes the triangle of the icosahedron and the pentagon of the dodecahedron inscribed in the same sphere,'[15] which is the theorem of Aristaeus, and expressed, moreover, in nearly the same words as in Hypsicles;

2. Pappus says in Book VII., as we have seen, pp. 194-5, that the works in the τόπος ἀναλυόμενος—of which the τόποι στερεοί of Aristaeus is one—proceed by the method of analysis and synthesis; and it is to be observed that the investigation in Pappus of the problem, 'to inscribe the regular solids,' is made by the analytical method;[16]

3. Pappus, moreover, in Book V., treats of 'the comparison of the five figures having equal surface, viz. the pyramid, cube, octahedron, dodecahedron and icosahedron,' and says that he will do so, 'not by the so called analytic method, by which some of the ancients (τῶν παλαιῶν) found their proofs, but by the synthetic method arranged by him in a more perspicuous and shorter manner'—ἑξῆς δὲ τούτοις γράψομεν, ὡς ὑπεσχόμεθα, τὰς συγκρίσεις τῶν ἴσην ἐπιφάνειαν ἐχόντων πέντε σχημάτων, πυραμίδος τε καὶ κύβου καὶ ὀκταέδρου δωδεκαέδρου τε καὶ εἰκοσαέδρου, οὐ διὰ τῆς ἀναλυτικῆς λεγομένης θεωρίας, δι' ἧς ἔνιοι τῶν παλαιῶν ἐποιοῦντο τὰς ἀποδείξεις, ἀλλὰ διὰ τῆς κατὰ σύνθεσιν ἀγωγῆς ἐπὶ τὸ σαφέστερον καὶ συντομώτερον ὑπ' ἐμοῦ διεσκευασμένας.[17]

The *theorem of Aristaeus* can be proved in the following simple manner:—

[14] *L'Arithmétique des Grecs dans Pappus*, Mémoires de la Société des Sciences phys. et nat. de Bordeaux, 2e Série, tom. III., p. 351. 1880.

[15] Pappi *Collect.*, vol. I., p. 162.

[16] *Ibid.*, pp. 142-162.

[17] *Ibid.*, pp. 410, 412.

If a regular dodecahedron be inscribed in a sphere, the poles of its faces will be the vertices of a regular icosahedron inscribed in the same sphere; and, conversely, the vertices of the dodecahedron will be the poles of the faces of the icosahedron. Now let R be the pole of the circle circumscribing the pentagon $ABCDE$ of the dodecahedron, and let S and T be the poles of the circles circumscribing the two other pentagons of the dodecahedron which have the vertex A in common: then A will be the pole of the circle circumscribing the triangle RST of the icosahedron. Now, if the points R and A be joined to O, the centre of the sphere, the lines OR, OA so drawn will be at right angles to the planes $ABCDE$ and RST respectively: let them intersect these planes at the points P and Q respectively. Then the two right-angled triangles ORQ, OAP—having equal hypotenuses OR, OA, and common angle ROA—will be equal in every respect; therefore $OP = OQ$ and $AP = BQ$. But AP and BQ are the radii of the circles circumscribing the pentagon of the dodecahedron and the triangle of the icosahedron, and OP, OQ are the perpendiculars drawn from the centre to these two planes.

In the second chapter of this work (p. 38, *sq.*), we saw that 'the Pythagoreans were much occupied with the construction of regular polygons and solids, which in their cosmology played an essential part as the fundamental forms of the elements of the universe':[18] and in the third

[18] These Pythagorean ideas—which were adopted by Plato, Πλάτων δὲ καὶ ἐν τούτοις πυθαγορίζει (see p. 86, *n.* 76)—played such an important part in antiquity that they gave rise to the belief, related by Proclus, that Euclid 'proposed to himself the construction of the so-called Platonic bodies [the regular solids] as the final aim of his systematisation of the Elements' (p. 6). This has been noticed by P. Ramus, who says: 'Nihil in antiqua geometria speciosius visum est quinque corporibus ordinatis, eorumque gratia geometriam ut ex Proclo initio dictum est, inventam esse veteres illi crediderunt'; but he adds: 'At in totis elementis nihil est istis argutiis ineptius et inutilius.'*

* (Petri Rami *Scholarum Mathematicarum* Libri unus et triginta. Francofurti, 1599, p. 306.)

chapter (p. 86, *sq.*), I pointed out a problem of high philosophical importance to the Pythagoreans, which, in my judgment, naturally arose from their cosmological speculations, and which required for its solution a knowledge of stereometry, and also the solution of the famous problem: *to find two mean proportionals between two given lines.* In the same chapter (p. 88) I indicated the men who first solved this problem, and laid the foundation of stereometry: in the following chapters I examined their work; and finally in this chapter we have seen that Aristaeus wrote works on the conic sections and on the regular solids, and, further, that he is specially mentioned as one of those who cultivated the analytic method—the method by the aid of which these discoveries were made, as stated in p. 88. Aristaeus may, therefore, be regarded as having continued and summed up the work, which, arising from the speculations of Philolaus, was carried on by his successors—Archytas, Eudoxus, and Menaechmus. These men were related to one another in succession as master and pupil, and it seemed to me important that the continuity of their work should not be broken in its presentation.

[It may be interesting to some of the readers of this work to know that William Allman, M.D., Professor of Botany in the University of Dublin (1809–1844), and father of the writer, in a Memoir entitled: 'An attempt to Illustrate a Mathematical Connection between the parts of Vegetables' (read before the Royal Society of London in the year 1811), put forward the hypothesis that the minute cells in the young shoots of vegetables are of the dodecahedral form in Dicotyledonous plants; and of the icosahedral form in Monocotyledonous plants; and that by means of this hypothesis he accounted for the prevalence of the number 5, and the exogenous growth in the former, and of the number 3, and the endogenous growth in the latter.]

CHAPTER IX.*

THEAETETUS.

Theaetetus of Athens.—Notices of his work.—Passage from Plato's 'Theaetetus' quoted and annotated.—Theaetetus first wrote on the five 'Regular Solids.'—In the composition of his 'Elements' Euclid was most indebted to the Pythagoreans, Theaetetus and Eudoxus.—What portions of the 'Elements' are due to each of these sources?—The principal part of the original work of Euclid himself is contained in the Tenth Book.—Probable object of this Book.—Pythagoras discovered the Theory of Incommensurables, and it is probable that the Pythagoreans went farther in this research than is commonly supposed.—Conclusion.

AT the close of the last chapter I pointed out the connection between the several parts of this work, and stated the reasons for the order which I followed. This order was founded on the belief that the true history of Greek geometry was most correctly represented by exhibiting in an unbroken series the work done by Archytas and his successors. This course of proceeding led to the temporary omission of at least one geometer, who had greatly advanced the science.

THEAETETUS of Athens, a pupil of Theodorus of Cyrene, and also a disciple of Socrates, is represented by Plato, in the dialogue which bears his name, as having impressed both his teachers by his great natural gifts and genius.

* Within the last year the following works have been published: Euclidis *Elementa*, edidit et Latine interpretatus est J. L. Heiberg, Dr. Phil., vol. III., librum X. continens, Lipsiae, 1886; *die Lehre von den Kegelschnitten im Altertum*, von Dr. H. G. Zeuthen, zweiter halbband, Kopenhagen, 1886; *Notice sur les deux Lettres Arithmétiques de Nicolas Rhabdas* (texte Grec et traduction), par M. Paul Tannery (Extrait des notices et extraits des manuscrits de la Bibliothèque nationale, &c., tom. XXXII., 1re Partie), Paris, 1886.

A new journal, devoted to *the History of Mathematics*, has been founded this year by Dr. Gustaf Eneström, of Stockholm:—*Bibliotheca Mathematica*, Journal d'Histoire des Mathematiques.

All that we know of his work is contained in the following notices :—

(*a*). He is mentioned by Eudemus in the passage quoted from Proclus in the *Introduction* (p. 4), along with his contemporaries Archytas of Tarentum, and Leodamas of Thasos, as having increased the number of demonstrations of theorems and solutions of problems, and developed them into a larger and more systematic body of knowledge;[1]

(*b*). We learn from the same source that Hermotimus of Colophon advanced yet further the stores of knowledge acquired by Eudoxus and Theaetetus, and that he discovered much of the 'Elements,' and wrote some parts of the 'Loci';[2]

(*c*). Proclus, speaking of the collection of the 'Elements' made by Euclid, says that he arranged many works of Eudoxus, and completed many of those of Theaetetus;[3]

(*d*). The theorem Euclid X. 9 :—'The squares on right lines, commensurable in length, have to each other the ratio which a square number has to a square number; and conversely. But the squares on right lines incommensurable in length have not to each other the ratio which a square number has to a square number; and conversely'—is attributed to Theaetetus by an anonymous Scholiast, probably Proclus. The scholium is :—*τοῦτο τὸ θεώρημα Θεαιτήτειόν ἐστιν εὕρημα καὶ μέμνηται αὐτοῦ Πλάτων ἐν Θεαιτήτῳ, ἀλλ' ἐκεῖ μὲν μερικώτερον ἔγκειται* [*ἔκκειται*], *ἐνταῦθα δὲ καθόλου*;[4]

(*e*). In the passage referred to, Theaetetus relates how his master Theodorus—who was subsequently the mathematical teacher of Plato—had been writing out for him

[1] Proclus, ed. Friedlein, p. 66.

[2] *Ibid.*, p. 67.

[3] *Ibid.*, p. 68.

[4] Knoche, *op. cit.*, p. 24; cf. F. Commandinus, *Euclidis Elementorum Libri* XV., *una cum Scholiis antiquis*, fol. 129, p. 2, Pisauri, 1619.

and the younger Socrates something about squares:[5] about the squares whose areas are three feet and five feet, showing that in length they are not commensurable with the square whose area is one foot[6] [that the sides of the squares whose areas are three superficial feet and five superficial feet are incommensurable with the side of the square whose area is the unit of surface, *i. e.* are incommensurable with the unit of length], and that Theodorus had taken up separately each square as far as that whose area is seventeen square feet, and, somehow, stopped there. Theaetetus continues:—'Then this sort of thing occurred to us, since the squares appear to be infinite in number,[7]

[5] Περὶ δυνάμεών τι ἡμῖν Θεόδωρος ὅδε ἔγραφε, τῆς τε τρίποδος πέρι καὶ πεντέποδος ἀποφαίνων ὅτι μήκει οὐ ξύμμετροι τῇ ποδιαίᾳ. In mathematical language δύναμις signifies 'power,' especially the second power or square. In the passage (*e*), however, the word seems not to be used steadily in the same signification, and in 148 A it certainly means 'root.' M. Paul Tannery considers that the present text of Plato is corrupt, and that in it δύναμις (power) should be replaced throughout by δυναμένη (root). Professor Campbell (*Theaetetus of Plato*, p. 21, note) thinks that 'it is not clear that in Plato's time this point of terminology was fixed.' But, on the other hand, J. Barthélemy Saint-Hilaire believes that the expression, δύναμις, was probably invented by the Pythagoreans (*Métaphysique d'Aristote*, tom. II., p. 156, *n.* 16). In support of this view it may be noticed that the term δυνάμει is used in its proper signification throughout the oldest fragment of Greek geometry—that handed down by Simplicius from the *History of Geometry* of Eudemus on the quadrature of the lunes (see pp. 69–75; and, for the revised Greek text, Simplicii *in Aristotelis Physicorum libros quatuor priores commentaria*, ed. H. Diels, pp. 61–68, Berlin, 1882)—and is so used, for the most part, in paragraphs which, according to the *criterion* laid down in p. 72, *n.* 45, must be regarded as genuine. Now since Eudemus, in this fragment, gives an analysis of the work of Hippocrates, and, moreover, frequently refers to him by name, it is probable that, in parts at least, he quoted the work on lunes textually, and that the word δυνάμει, which occurs throughout, must have been used by Hippocrates, who we know was connected with the Pythagoreans. On the whole then it seems to me probable that Plato had not fully grasped the distinction between the terms δύναμις and δυναμένη; and that in this is to be found the true explanation of the obscurity of the passage.

[6] μήκει οὐ ξύμμετροι τῇ ποδιαίᾳ. See Euclid X., *Def.* 1. Σύμμετρα μεγέθη λέγεται τὰ τῷ αὐτῷ μέτρῳ μετρούμενα, ἀσύμμετρα δέ, ὧν μηδὲν ἐνδέχεται κοινὸν μέτρον γενέσθαι. 2. Εὐθεῖαι δυνάμει σύμμετροί εἰσιν, ὅταν τὰ ἀπ' αὐτῶν τετράγωνα τῷ αὐτῷ χωρίῳ μετρῆται, ἀσύμμετροι δέ, ὅταν τοῖς ἀπ' αὐτῶν τετραγώνοις μηδὲν ἐνδέχηται χωρίον κοινὸν μέτρον γενέσθαι.

[7] ἐπειδὴ ἄπειροι τὸ πλῆθος αἱ δυνάμεις ἐφαίνοντο. Cf. Eucl. X., *Def.* 3: τούτων

to try and comprise them in one term, by which to designate all these squares.'

Socr. 'Did you discover anything of the kind?'

Theaet. 'In my opinion we did. Attend, and see whether you agree.'

Socr. 'Go on.'

Theaet. 'We divided all number into two classes: comparing that number which can be produced by the multiplication of equal numbers to a square in form, we called it quadrangular and equilateral.'[8]

Socr. 'Very good.'

Theaet. 'The numbers which lie between these, such as three and five, and every number which cannot be produced by the multiplication of equal numbers, but becomes either a larger number taken a lesser number of times, or a lesser taken a greater number of times (for a greater factor and a less always compose its sides); this we likened to an oblong figure, and called it an oblong number (προμήκη ἀριθμόν).'[9]

ὑποκειμένων δείκνυται, ὅτι τῇ προτεθείσῃ εὐθείᾳ ὑπάρχουσιν εὐθεῖαι πλήθει ἄπειροι σύμμετροί τε καὶ ἀσύμμετροι αἱ μὲν μήκει μόνον, αἱ δὲ καὶ δυνάμει.

[8] τὸν ἀριθμὸν πάντα δίχα διελάβομεν. τὸν μὲν δυνάμενον ἴσον ἰσάκις γίγνεσθαι τῷ τετραγώνῳ τὸ σχῆμα ἀπεικάσαντες τετράγωνόν τε καὶ ἰσόπλευρον προσείπομεν. Cf. Euclid, VII., *Def.* 19: τετράγωνος ἀριθμός ἐστιν ὁ ἰσάκις ἴσος ἢ [ὁ] ὑπὸ δύο ἴσων ἀριθμῶν περιεχόμενος; also Aristotle, *Anal. Post.* I. iv., οἷον τὸ εὐθὺ ὑπάρχει γραμμῇ καὶ τὸ περιφερές, καὶ τὸ περιττὸν καὶ ἄρτιον ἀριθμῷ, καὶ τὸ πρῶτον καὶ σύνθετον καὶ ἰσόπλευρον καὶ ἑτερόμηκες (see Euclid, VII., *Def.* 7, 6, 12, 14). Plato's expression is tautologous.

[9] τὸν τοίνυν μεταξὺ τούτου, ὧν καὶ τὰ τρία καὶ τὰ πέντε καὶ πᾶς ὃς ἀδύνατος ἴσος ἰσάκις γενέσθαι, ἀλλ' ἢ πλείων ἐλαττονάκις ἢ ἐλάττων πλεονάκις γίγνεται, μείζων δὲ καὶ ἐλάττων ἀεὶ πλευρὰ αὐτὸν περιλαμβάνει, τῷ προμήκει αὖ σχήματι ἀπεικάσαντες προμήκη ἀριθμὸν ἐκαλέσαμεν. Cf. Euclid, VII., *Def.* 17: Ὅταν δὲ δύο ἀριθμοὶ πολλαπλασιάσαντες ἀλλήλους ποιῶσί τινα, ὁ γενόμενος ἐπίπεδος καλεῖται, πλευραὶ δὲ αὐτοῦ οἱ πολλαπλασιάσαντες ἀλλήλους ἀριθμοί. From the time of Pythagoras—to whom the combination of arithmetic with geometry was due—the properties of numbers were investigated geometrically. Thus composite numbers (σύνθετοι) were figured as rectangles, whose sides (πλευραί) are the factors. Similarly, prime numbers (πρῶτοι) were represented by points ranged along a right line, and were hence called linear (γραμμικοί) not only by Theon. of Smyrna (*Arithm.* ed. de Gelder, p. 34), and Nicomachus (Nicom. G., *Introd. Arithm.*, II. c. 7), but also by Speusippus, who wrote a little work *on Pythagorean numbers*

Socr. 'Capital! What next?'

Theaet. 'The lines which form as their squares an equilateral plane [square] number we defined as μῆκος [length, *i. e.* containing a certain number of linear units], and the lines which form as their squares an oblong number (τὸν ἑτερομήκη) we defined as δυνάμεις, inasmuch as they have no common measure with the former in length, but in the surfaces of the squares, which are equivalent to these oblong numbers. And in like manner with solid numbers.' [10]

Socr. 'The best thing you could do, my boys; no one could do better.'—(*Theaetetus*, 147 D–148 B.)

(see *Theologumena Arithmetica*, ed. Ast., p. 61). Prime numbers were also figured as rectangles whose common breadth was the linear unit, and they are thus represented in this passage.

In geometry τὸ ἑτερόμηκες signified a rectangle, and was so defined by Euclid, Book I., *Def.* 22: τῶν δὲ τετραπλεύρων σχημάτων τετράγωνον μὲν ἔστιν, ὃ ἰσόπλευρόν τέ ἐστι καὶ ὀρθογώνιον, ἑτερόμηκες δὲ, ὃ ὀρθογώνιον μέν, οὐκ ἰσόπλευρον δέ. Cf. Hero, *Def.* 53; *Geom.*, pp. 43, 52, 53, &c., ed. Hultsch; Pappi Alex. *Collect.*, ed. Hultsch, vol. I., p. 140. Euclid does not use the term ἑτερόμηκες in his *Elements*, but παραλληλόγραμμον ὀρθογώνιον. It is now generally recognised that he derived the materials of his *Elements* from various sources: the term ἑτερόμηκες may thus have been preserved in his work; or, else, he thought it better to avoid the use of this term, as it was employed in a particular sense. When the sides of the rectangle were expressed in numbers, προμήκης was the general name for an oblong. In the particular cases where the sides of the oblong contained two consecutive units, as—2, 3; 3, 4; &c., the term ἑτερομήκης was employed, inasmuch as the lengths of the sides were of different kinds, *i. e.* odd and even; whereas in a square they were of the same kind, either both odd, or both even (see the second chapter, p. 32, *n.* 51). It should be observed that when a square is constructed equal to an oblong of this kind (ἑτερόμηκες), its side must be incommensurable; but in certain cases the side of the square, which is equal to an oblong of the former kind (πρόμηκες) (*e. g.* whose sides are 8, 2; 3, 27; and so on) is commensurable. The two words are used in this passage in their strict signification, and are not, as M. Paul Tannery thinks, synonymous (see *Domninos de Larissa*, Bulletin des Sciences math. et astrom., tom. VIII., 1884, p. 297). Professor Campbell remarks: 'these terms [προμήκης ἑτερομήκης] were distinguished by the later Pythagoreans' (*loc. cit.*, p. 23, *n.*). This is misleading, for it seems to imply that they were not distinguished by the early Pythagoreans.

[10] ὅσαι μὲν γραμμαὶ τὸν ἰσόπλευρον καὶ ἐπίπεδον ἀριθμὸν τετραγωνίζουσι, μῆκος ὡρισάμεθα, ὅσαι δὲ τὸν ἑτερομήκη, δυνάμεις, ὡς μήκει μὲν οὐ ξυμμέτρους ἐκείναις, τοῖς δ' ἐπιπέδοις ἃ δύνανται· καὶ περὶ τὰ στερεὰ ἄλλο τοιοῦτον. Cf. Euclid, VII., *Def.* 18: ὅταν δὲ τρεῖς ἀριθμοὶ πολλαπλασιάσαντες ἀλλήλους ποιῶσί τινα, ὁ γενόμενος στερεός ἐστιν, πλευραὶ δὲ αὐτοῦ οἱ πολλαπλασιάσαντες ἀλλήλους ἀριθμοί.

(*f*). We learn from Suidas that he taught at Heraclea, and that he first wrote on 'the five solids' as they are called.[11]

Eudoxus and Theaetetus, then, were the original thinkers to whom—after the Pythagoreans—Euclid was most indebted in the composition of his 'Elements.' In the former chapters of this work we have seen that we owe to the Pythagoreans the substance of the first, second, and fourth Books, also the doctrine of proportion and of the similarity of figures, together with the discoveries respecting the *application*, *excess*, and *defect* of areas[12]—the subject matter of the Sixth Book: the theorems arrived at, however, were proved for commensurable magnitudes only, and assumed to hold good for all. We have seen, further, that the doctrine of proportion, treated in a general manner, so as to include incommensurables (Book V.), and, consequently, the recasting of Book VI., and also the Method of Exhaustions (Book XII.), were the work of Eudoxus. If we are asked now—In what portion of the Elements does the work of Theaetetus survive? We answer: since Books VII., VIII., and IX. treat of numbers, and our question concerns geometry; and since the substance of Book XI., containing, as it does, the basis of the geometry of volumes, is probably of ancient date, we are led to seek for the work

Solid numbers (στερεοί) were also treated in the little work of Speusippus referred to above (*Theol. Arith. loc. cit.*).

[11] 'Theaetetus, of Athens, astronomer, philosopher, disciple of Socrates, taught at Heraclea. He first wrote on "the five solids" as they are called. He lived after the Peloponnesian war.'

'Theaetetus, of Heraclea in Pontus, philosopher, a pupil of Plato.' *Sub. v.*

It has been conjectured that the two Notices refer to the same person. Making every allowance for the inaccuracy of Suidas, this seems to me by no means probable. It is much more likely that the second was a son, or relative, of Theaetetus of Athens, and sent by him to his native city to study at the Academy under Plato.

[12] By this method the Pythagoreans solved geometrical problems, which depend on the solution of quadratic equations. For examples of the method, see *supra*, p. 41, and p. 72, *n.* 46.

of Theaetetus in Books X. and XIII.: and it is precisely with the subjects of these Books that the extracts (*d*), (*e*), and (*f*) are concerned.

Having regard, however, to the difference in the manner of expression of Proclus in (*c*) :—'Euclid *arranged* many works of Eudoxus, and *completed* many of those of Theaetetus'—we infer that, whereas the bulk of the fifth and twelfth Books are due to Eudoxus, on the other hand Theaetetus laid the foundation only of the doctrine of incommensurables, as treated in the tenth Book. In like manner, from (*f*) we infer that the thirteenth Book, treating of the regular solids, is based on the theorems discovered by Theaetetus; but it contains, probably, 'a recapitulation, at least partial, of the work of Aristaeus' (p. 202).

From what precedes, it follows that the principal part of the original work of Euclid himself, as distinguished from that of his predecessors, is to be found in the Tenth Book.[13] De Morgan suspected that in this Book some definite object was sought, and suggested that the classification of incommensurable quantities contained in it was undertaken in the hope of determining thereby the ratio of the circumference of the circle to its diameter, and thus solving the vexed question of its quadrature.[14] It is

[13] See Heiberg., *Studien über Euklid*, p. 34: 'Nach Proklus hat er [Euklid] vieles von den Untersuchungen des Theätet vervollkommnet; also, da Theätet sich besonders mit Inkommensurabilität und Irrationalität beschäftigte, darf wohl einiges von dem sehr umfangreichen und vollständigen X Buche dem Euklid selbst angeeignet werden, was und wie viel, wissen wir nicht.'

Professor P. Mansion, of the University of Ghent, informs me by a letter of the 4th March, 1887, that for several years past he has pointed out this result—the originality of the Tenth Book of the Elements of Euclid—to his pupils in his Course on the History of Mathematics. His manner of proof is substantially the same as that given by me above.

See also M. Paul Tannery: *l'Éducation Platonicienne*, Revue Philosophique, Mars, 1881, p. 295; *la Constitution des Éléments*, Bulletin des Sciences math. et astron., 1886, p. 190.

[14] The English Cyclopaedia, *Geometry*, vol. IV., 375; Smith's Dictionary of Greek and Roman Biography, *Eucleides*, vol. II., p. 67.

more probable, however, that the object proposed concerned rather the subject of Book XIII., and had reference to the determination of the ratios between the edges of the regular solids and the radius of the circumscribed sphere, ratios which in all cases are irrational.[15] In this way is seen, on the one hand, the connection which exists between the two parts of the work of Theaetetus, and, on the other, light is thrown on the tradition handed down by Proclus, and referred to at the end of the last chapter, that 'Euclid proposed to himself the construction of the so-called Platonic bodies [the regular solids] as the final aim of his systematisation of the Elements.'

We are not justified in inferring from the passage in *Theaetetus* (*e*), that Theodorus had written a work on 'powers' or 'roots,' much less that the contribution of the Pythagoreans to the doctrine of incommensurables was limited to proving the incommensurability of the diagonal and side of a square, *i. e.* of $\sqrt{2}$.[16] Theodorus, who was a teacher of mathematics, is represented in the passage merely as showing his pupils the incommensurability of $\sqrt{3}, \sqrt{5}, \ldots \sqrt{17}$, and there is no evidence that this work was original on his part. On the contrary, the knowledge of the incommensurability of $\sqrt{5}$, at all events, must be attributed to the Pythagoreans, inasmuch as it is an immediate consequence of the incommensurability of the segments of a line cut in extreme and mean ratio which must have been known to them, and from which indeed it is probable that the existence of incommensurable lines was discovered by Pythagoras himself (see *supra*, p. 42, and pp. 137–8).

There are, moreover, good reasons for believing that the Pythagoreans went farther in this research than has been sometimes supposed; indeed Eudemus says ex-

15 See Bretsch., *Geom. vor Eukl.*, p. 148.

16 See P. Tannery, *op. cit.*, pp. 188, 189.

pressly: 'Pythagoras discovered the theory of incommensurable quantities (τῶν ἀλόγων πραγματείαν).' Further, the lines $\sqrt{3}$, $\sqrt{5}$, . . . would occur in many investigations with which we know the Pythagoreans were occupied:—

1°. In the endeavour to find the so-called *Pythagorean triangles*, i. e. right-angled triangles in rational numbers;

2°. In the determination of a square, which shall be any multiple of the square on the linear unit, a problem which can be easily solved by successive applications of the 'Theorem of Pythagoras'—the first right-angled triangle, in the construction, being isosceles, whose equal sides are the linear unit; the second having for sides about the right angle the hypotenuse of the first ($\sqrt{2}$) and the linear unit; the third having for sides about the right angle $\sqrt{3}$ and 1, and for hypotenuse 2, and so on;

3°. In the construction of the regular polygons, for the third triangle in 2° is, in fact, the so-called 'most beautiful right-angled scalene triangle' (p. 38);

4°. In finding a mean proportional between two given lines, or the construction of a square which shall be equal to a given rectangle, in the simple case when one line is the linear unit, and the other contains 3, 5, . . . units.

The method followed in this work differs altogether from that pursued by most writers. The usual course has been to treat of the works of Archytas, Theaetetus, Eudoxus, Menaechmus, &c.—the men to whom in fact, as we have seen, the progress of geometry at that time was really due—under the head of 'Plato and the Academy.' This has given rise to an exaggerated view of the services of Plato and of the Academy in the advancement of mathematics: which is the more remarkable because a just appreciation of the services of Plato in this respect was made by Eudemus in the summary of the history of geometry, so frequently quoted in these pages:

'Plato, who came next after them [Hippocrates of Chios, and Theodorus of Cyrene], caused the other branches of knowledge to make a very great advance through his earnest zeal about them, and especially geometry: it is very remarkable how he crams his essays throughout with mathematical terms and illustrations, and everywhere tries to rouse an admiration for them in those who embrace the study of philosophy.'[17]

The way in which Plato is here spoken of is in striking contrast to that in which Eudemus has, in the summary, written of the promoters of geometry.

[17] Πλάτων δ' ἐπὶ τούτοις γενόμενος, μεγίστην ἐποίησεν ἐπίδοσιν τά τε ἄλλα μαθήματα καὶ τὴν γεωμετρίαν λαβεῖν διὰ τὴν περὶ αὐτὰ σπουδήν, ὅς που δῆλός ἐστι καὶ τὰ συγγράμματα τοῖς μαθηματικοῖς λόγοις καταπυκνώσας καὶ πανταχοῦ τὸ περὶ αὐτὰ θαῦμα τῶν φιλοσοφίας ἀντεχομένων ἐπεγείρων. Proclus, ed. Friedlein, p. 66.

NOTES AND ADDITIONS.

Continuation of Bibliographical Notices.*

SINCE the publication of the concluding part of this work in *Hermathena* (July, 1887), M. Paul Tannery has collected his Papers, which appeared in the *Bulletin des Sciences mathématiques et astronomiques*, since April, 1885, and published them in a volume entitled: *La Géométrie Grecque comment son histoire nous est parvenue et ce que nous en savons. Essai Critique.* Première partie. *Histoire générale de la géométrie élémentaire.* Paris, 1887.

M. Paul Tannery has also published a volume on the origin of science in general—*Pour l'Histoire de la Science Hellène de Thalès à Empédocle.* Paris, 1887. This work is founded on articles which were published by M. Tannery in the *Revue philosophique.*

Dr. Heiberg has completed his edition of the *Elements* of Euclid by the publication of vol. v.—*Continens Elementornm qui feruntur Libros XIV.–XV. et Scholia in Elementa cum Prolegomenis criticis et Appendicibus.* Lipsiae, 1888.

The first part of a Monograph on Eudoxus by Herr Hans Künsberg has recently appeared—*Der Astronom, Mathematiker und Geograph EUDOXOS, von Knidos,* I. Theil: *Lebensbeschreibung des Eudoxos, Ueberblick über seine astronomische Lehre und geometrische Betrachtung der Hippopede* von Hans Künsberg, kgl. Reallehrer. (Programm zum Jahresbericht der vierkursigen königl. Realschule Dinkelsbühl pro 1888.) Druck von C. Fritz in Dinkelsbühl.

There has also been recently published: *A Short Account of the History of Mathematics,* by Walter W. Rouse Ball, Fellow and Assistant-Tutor of Trinity College, Cambridge; and of the Inner Temple, Barrister-at-law. London, Macmillan and Co., 1888. This book is for the most part a transcript of some lectures delivered this year by Mr. W. W. Rouse Ball.

* See pp. 1, 52, 150, 180, and 206.

Pages 11, 12.

The passage of Geminus referred to here is taken from his *Review of Mathematics*, and is given *in extenso* in Chapter VI., pp. 164, 165.

Pages 16, 80.

Harpedonaptae. See Cantor (*Vorlesungen über Geschichte der Mathematik*, pp. 55–57), who points out the Greek origin (ἁρπεδόνη, a rope, and ἅπτειν, to fasten), previously overlooked, of this name, and shows from inscriptions on the Egyptian temples that the duty of these 'rope-fasteners' consisted in the orientation of the buildings by reference to the constellation of the Great Bear. The meridian being thus found, the line at right angles to it was probably determined by the construction of a triangle with ropes measuring 3, 4 and 5 lengths respectively. We have seen (p. 29) that the Egyptians knew that such a triangle would be right-angled. The operation of rope-stretching, Cantor adds, was one of unknown antiquity, being noticed in a record of the time of Amenemhat I., which is preserved in the Berlin Museum.

Pages 29–32.

In connection with this passage of Plutarch, and the observations thereon, it is interesting to note that M. Paul Tannery (*la Géométrie Grecque*, p. 105) has found in G. Pachymeres (MSS. de la Bibliothèque nationale) the expression τὸ θεώρημα τῆς νύμφης, to designate the 'theorem of Pythagoras' (Euclid I. 47). In a letter to me, of July 3, 1886, M. Tannery mentions that the Arabs call it 'the theorem of the bride.' This name for the theorem seems to point to the old Egyptian idea as handed down by Plutarch.

PAGE 37, NOTE 58.

I have since found in Billingsley's *Euclide*[1] the following note on I. 43:—

'This proposition Pelitarius calleth Gnomicall, and misticall for that of it (sayeth he) spring infinite demonstrations and uses in Geometry.' (Fol. 54.)

On referring to Peletarius, however, it will be found that he only calls the figure *Gnomic:* not the proposition. After the demonstration of I. 43, he says: 'Vix enim usquam in toto opere Geometrico occurrit Figuratio magis fœcunda quam haec Gnomica: hoc est, quae uno parallelogrammo et Gnoma conflatur . . . Nam hîc Gnomonis explicandi locus est maximè oportunus: licet Euclides ad secundum librum distulerit.

'Hanc ego Figuram mysticam soleo vocare: Ex ea enim, velut ex locupletissimo promptuario, innumerabiles exeunt Demonstrationes. Quod cum magna voluptate perspiciet qui re Geometrica seriò se exercebit.' (Iacobi Peletarii Cenomani, *in Euclidis Elementa Geometrica Demonstrationum Libri Sex*, p. 41. Lugduni, apud Ioan. Tornæsium et Gul. Gazeium, 1557.)

PAGE 43, NOTE 64.

This proposition—Euclid X. 117—is an interpolation, and is recognised as such by August, who gives it in Appendix V., pars. ii.,

[1] *The Elements of Geometrie of the most auncient Philosopher EUCLIDE of Megara. Faithfully (now first) translated into the Englishe toung,* by H. Billingsley, Citizen of London.

Whereunto are annexed certaine Scholies, Annotations, and Inventions, of the best Mathematiciens, both of time past, and in this our age.

With a very fruitfull Præface made by M. I. Dee, specifying the chiefe Mathematicall Scieces, what they are, and whereunto commodious: where, also, are disclosed certaine new Secrets Mathematicall and Mechanicall, untill these our daies, greatly missed.

Imprinted at London by John Daye, 1570.

De Morgan (Smith's Dictionary of Greek and Roman Biography, *Eucleides*, vol. II., p. 73) states that Henry Billingsley "was a rich citizen and was mayor (with knighthood) in 1591."

p. 296; and by Heiberg, who gives it in the Appendix to Book X., (Euclidis *Elementa*, vol. III., p. 408). In Billingsley's *Euclide*, after this theorem, which is prop. 116 in that edition, 'An other demonstration after *Flussas*' is given. Then follows the observation: 'This demonstration I thought good to adde, for that the former demonstrations seme not so full, and they are thought of some to be none of *Theons*, as also the proposition to be none of *Euclides*.' (Fol. 310, p. 2.)

PAGES 49 AND 132 (d).

P. Ramus suspected that this Scholium was due to Proclus; he says: 'Quintum librum Scholiastes graecus Arcadius nempe vel Pappus vel quod apparet é 19 p. 10 [Euclid X. 19],[1] Proclus refert ad Eudoxum Platonis praeceptorem, quem tamen Proclus sodalem Platonis efficit, et scopum ait esse libri de analogiis, et certé de solis analogiis agitur libro quinto. Proclus putat totum librum hunc esse communem Arithmeticae et Geometriae . . . . . .' (Petri Rami *Scholarum Mathematicarum* libri unus et triginta, p. 212 Francofurti, 1599.)

PAGE 59.

Last line 'Αριστοτελικὰ Κηρία, the name of a collection made by Porus (Sporus) of Nicaea. See Archimedis *Opera*, ed. Heiberg, vol. III., pp. 264 and 300: cf. *infra* pp. 184–5. Dr. Heiberg, however, distinguishes the κηρία of Porus from the κηρία of Aristotle, and thinks that by the latter is meant his treatise περὶ σοφιστικῶν ἐλέγχων (*loc. cit.*, p. 265).

[1] See Commandinus, *Euclidis Elementorum, Libri XV., una cum Scholiis antiquis*, fol. 135, p. 2, Pisauri, 1619; also Euclidis *Elem.* Graece ed. ab August, pars ii., p. 284; and Billingsley's *Euclide*, fol. 246, in each of which a certain lemma, scholium, or annotacion, is attributed to Proclus.

PAGE 61.

M. Paul Tannery (*la Géométrie Grecque*, p. 81) thinks that this passage of Iamblichus has hitherto been misunderstood. The last sentence is translated by M. Tannery thus: 'Voici comment les Pythagoriciens disent que la Géométrie fut rendue publique. L'argent des Pythagoriciens fut perdu par l'un d'eux;*[1] à la suite de ce malheur, on lui accorda de battre monnaie avec la Géométrie, —et la Géométrie fut appelée *Tradition touchant Pythagore*.'[2] M. Tannery infers that the last words of the passage were the title of a work on Geometry which Eudemus had in his possession, and from which he derived his information concerning the works of the Pythagorean School.

I am unable to agree with M. Tannery either as to the interpretation of this passage or in the inference he draws from it.

PAGES 64–75.

The first part of this extract—as far as p. 69—is taken by Simplicius chiefly from Alexander of Aphrodisias, and the remainder—from p. 69 to the end—from the second book of the *History of Geometry* of Eudemus.

The Aldine edition of the commentary of Simplicius on the *Physica Auscultatio* was published in 1526. In this edition the text of the fragment of Eudemus is admitted to be very inaccurate.

* He adds the following notes:—

'[1] Ἀποβαλεῖν τινα τὴν οὐσίαν τῶν Πυθαγορείων. On traduit d'ordinaire: "Un pythagoricien perdit sa fortune." Cette interprétation ne tient nullement compte de la construction de la phrase, ni des mœurs de l'époque à laquelle se rapporte la tradition. Les Pythagoriciens vivaient en communanté; le dépositaire de la bourse commune la perd, il faut recourir à des moyens extraordinaires. Voilà la légende; autrement elle ne se tient pas.

'[2] Ἐκαλεῖτο δὲ ἡ γεωμετρία πρὸς Πυθαγόρου ἱστορία, ce que Kiessling traduit: "Vocabatur autem Geometria a Pythagora *historia*." Il semble avoir entendu: "Pythagore appelait Géométrie *histoire*", interprétation insoutenable à tous les points de vue.'

M. Paul Tannery (*la Géométrie Grecque*) has noticed the successive attempts which have been made to explain and restore this fragment—a subject which he had treated more fully in the *Mémoires de la Société des Sciences physiques et naturelles de Bordeaux* (Tome v., 2e série, 1883). I give here the passage from the former work:—

'Bretschneider (1870) parvint le premier à expliquer convenablement l'extrait d'Eudème conservé par Simplicius, et par reconnaître qu'aucun paralogisme n'y est attribué à Hippocrate; qu'au contraire on trouve dans cet extrait une suite de théorèmes aussi intéressants qu' irréprochables.

'Quoique le document ne remonte pas à Hippocrate lui-même, il n'en serait pas moins inappréciable pour permettre de juger des connaissances géométriques de son époque, si malheureusement Simplicius, sous prétexte d'éclaircir un texte trop concis, ne s'était pas avisé d'y introduire des explications de son cru et de malencontreux développements, qui le défigurent singulièrement. La restitution du texte d'Eudème devient dès lors assez difficile pour que Bretschneider ait été entraîné à de graves erreurs, notamment à dénier à Hippocrate la connaissance de la propriété caractéristique des segments semblables, à savoir que tous les angles inscrits y sont égaux.

'M. Allman (*Hermathena*, iv., No. 7, p. 196–202; 1881) a, le premier, donné une traduction du texte d'Eudème, en le débarrassant des interpolations de Simplicius, d'après des règles dont l'application peut être discutée dans les détails, mais dont les principes sont hors de conteste. L'année suivante (Berlin, 1882) paraissait l'édition critique du Commentaire de Simplicius sur les quatre premiers Livres de la Physique d'Aristote, avec un texte singulièrement amélioré et un essai de distinction des interpolations dans le fragment d'Eudème (p. 61–68). Pour cette distinction, le savant éditeur, H. Diels, s'était aidé des lumières de M. Usener de Bonn, qui, en procédant suivant des principes analogues à ceux de M. Allman, est arrivé à des résultats concordants sur divers points, divergents sur d'autres. M. Diels a, d'autre part, inséré dans sa Préface, à la suite de remarques de M. Usener (p. XXIII.–XXVI.), quelques pages (XXVI.–XXXI.) d'observations critiques qu'il m'avait demandées, et dans lesquelles, tout en proposant des explications ou des corrections particulières pour certains passages obscurs, j'ai soutenu une partie des conclusions de M. Allman, en abandonnant les autres.

'J'ai repris depuis la question dans les *Mémoires de la Société des Sciences physiques et naturelles de Bordeaux* (v_2, p. 179–187: 1883), où j'ai publié le texte d'Eudème tel que je le comprenais, accompagné d'une traduction et des observations nécessaires. Enfin, M. Heiberg (*Philologus*, XLIII. 2, p. 337–344) a soumis ma restitution à une critique détaillée et proposé ses opinions sur divers points spéciaux.' (*la Géométrie Grecque*, pp. 116–7.)

PAGE 85, NOTE 73.

Biering, *Historia problematis cubi duplicandi*, Hauniae, 1884. Dr. Heiberg, in his notice of Cantor's 'History of Mathematics,' *Revue Critique d'Histoire et de Littérature*, 16 Mai, 1881, p. 380, remarks on this work: 'Je profite de l'occasion pour rappeler qu'il a été prouvé, lors de la soutenance orale de cette thèse, qu'elle n'est qu'un impudent plagiat de l'ouvrage de Reimer sur le même sujet.'

PAGES 93, 94.

With reference to the question discussed here concerning Hippias of Elis, see *infra*, pp. 189 *sq*.

TABLE

EXHIBITING THE CORRESPONDENCE BETWEEN THE SEVERAL PARTS OF THIS BOOK, AND THE ORIGINAL ARTICLES IN 'HERMATHENA.'

BOOK.	HERMATHENA.	DATE.
INTRODUCTION., . . CHAPTERS I., II., .	Vol. III., No. V., pp. 160–207, .	February, 1878.
CHAPTER III., . .	Vol. IV., No. VII., pp. 180–228, .	April, 1881.
CHAPTERS IV., V.,* .	Vol. V., No. X., pp. 186–235, .	May, 1884.
CHAPTER VI., . .	Vol. V., No. XI., pp. 403–432, .	July, 1885.
CHAPTERS VII., VIII.,	Vol. VI., No. XII., pp. 105–130, .	July, 1886.
CHAPTER IX., . . .	Vol. VI., No. XIII., pp. 269–278, .	July, 1887.

* The Bibliographical Notice at the end of Chapter V. appeared in *Hermathena*, Vol. V., No. XI.

INDEX.

THE END.

Printed in Dunstable, United Kingdom

91220017R00141